Genius Mathematician

Integral Private Class

Author: **Milad Hashemi**

Title: Genius Mathematician (Integral Private Class)

Author: Milad Hashemi

ISBN: 9781942912088

Publisher: Supreme Art, USA

Prepare for Publishing: Asan Nashr,

www.ASANASHR.com

Introduction of the author

After a few years of experience in solving various problems of integration and dialogue with different students, I realized that their main problem in solving problems of integration, lack of control over basic topics in mathematics. Therefore at the time of writing this book this discussion was my priority and wrote a different book. In fact by reading this book, in addition to integration, you will be familiarized with many other basic topics of mathematics, such as trigonometry, logarithm and exponential functions, absolute value function, the laws of exponents and radicals, differentiation, differential and... Also in the second chapter there are 95 unique problems with detailed responses through which there will be no need to refer to other sources. In fact, this book is like being in a private mathematics class that helps you to make progress in mathematics, especially integral discussion and I can assure you that after reading and mastery of this book, the integral discussion will no longer be difficult and unfamiliar to you. In the end I appreciate you for choosing this book for professional learning of mathematics particularly integral and sincerely thank you.

Thanks,

Milad Hashemi

The main differentiation formulas are as follows:

1) $y=a \rightarrow y'=0$

2) $y=x \rightarrow y'=1$

3) $y=ax \rightarrow y'=a$

4) $y=x^n \rightarrow y'=nx^{n-1}$

5) $y=u^n \rightarrow y'=nu^{n-1}u'$

6) $y=uv \rightarrow y'=u'v+v'u$

7) $y=\dfrac{u}{v} \rightarrow y'=\dfrac{u'v-v'u}{v^2}$

8) $y=\dfrac{1}{x} \xrightarrow{equal\ to} y=x^{-1} \rightarrow y'=-1x^{-2}=-\dfrac{1}{x^2}$

9) $y=\sqrt{x} \xrightarrow{equal\ to} y=x^{\frac{1}{2}} \rightarrow y'=\dfrac{1}{2}x^{-\frac{1}{2}} \xrightarrow{equal\ to} y'=\dfrac{1}{2}\times\dfrac{1}{x^{\frac{1}{2}}}=\dfrac{1}{2}\times\dfrac{1}{\sqrt{x}}=\dfrac{1}{2\sqrt{x}}$

10) $y=\dfrac{1}{u} \rightarrow y'=\dfrac{-u'}{u^2}$

11) $y=\sqrt{u} \rightarrow y'=\dfrac{u'}{2\sqrt{u}}$

12) $y=\sqrt[n]{u^m}=y=u^{\frac{m}{n}} \rightarrow y'=\dfrac{mu'}{n\sqrt[n]{u^{n-m}}}$

13) $y=|u| \rightarrow y'=\dfrac{uu'}{|u|}$

14) $y=\sin u \rightarrow y'=u'\cos u$

15) $y=\cos u \rightarrow y'=-u'\sin u$

16) $y=\tan u \rightarrow y'=u'(1+\tan^2 u)=\dfrac{u'}{\cos^2 u}=u'\sec^2 u$

17) $y=\cot u \rightarrow y'=-u'(1+\cot^2 u)=\dfrac{-u'}{\sin^2 u}=-u'\csc^2 u$

18) $y=\arcsin u \rightarrow y'=\dfrac{u'}{\sqrt{1-u^2}}$

19) $y=\text{arccos}u \rightarrow y'=\dfrac{-u'}{\sqrt{1-u^2}}$

20) $y=\text{arctan}u \rightarrow y'=\dfrac{u'}{1+u^2}$

21) $y=\text{arccot}u \rightarrow y'=\dfrac{-u'}{1+u^2}$

22) $y=\sinh u \rightarrow y'=u'\cosh u$

23) $y=\cosh u \rightarrow y'=u'\sinh u$

24) $y=\tanh u \rightarrow y'=u'(1-\tanh^2 u)$

25) $y=\coth u \rightarrow y'=u'(1-\coth^2 u)$

26) $y=\ln u \rightarrow y'=\dfrac{u'}{u}$

27) $y=\ln x \rightarrow y'=\dfrac{1}{x}$

28) $y=e^x \rightarrow y'=e^x$

29) $y=e^u \rightarrow y'=u'e^u$

30) $y=a^u \rightarrow y'=u'a^u\ln a$

Note: In Formula 1, 3 and 30 a is a fixed value

Note 2: Obviously u and v are functions of x.

The main integration formulas are as follows:

1) $\int x^n dx=\dfrac{x^{n+1}}{n+1}+c \overset{if}{\Rightarrow} n \neq -1$

2) $\int x^n \, dx = \ln|x| + c \overset{if}{\Rightarrow} n = -1$

3) $\int u^n \, du = \dfrac{u^{n+1}}{n+1} + c \overset{if}{\Rightarrow} n \neq -1$

4) $\int u^n du = \ln|u| + c \overset{if}{\Rightarrow} n=-1$

5)$\int sinu\ du = -cosu + c$

6)$\int cosu\ du = sinu + c$

7)$\int(1 + tan^2x)\ d\ x = \int \frac{1}{cos^2x} dx = \int sec^2x\ dx = tanx + c$

8)$\int(1 + cot^2 x)dx = \int \frac{1}{sin^2x} dx = \int csc^2 x\ dx = -cotx + c$

9)$\int secxtanx\ dx = secx + c$

10)$\int cscxcotx\ dx = -cscx + c$

11)$\int tanx\ dx = -ln|cosx| + c$

12)$\int cotx\ dx = ln|sinx| + c$

13)$\int \frac{1}{1+x^2} dx = arctanx + c$

14)$\int \frac{1}{1+u^2} du = arctanu + c$

15)$\int \frac{1}{\sqrt{1-x^2}} dx = arcsinx + c$

16)$\int \frac{1}{\sqrt{1-u^2}} du = arcsinu + c$

17)$\int \frac{1}{a^2+x^2} dx = \frac{1}{a} arctan\frac{x}{a} + c$

18)$\int \frac{1}{\sqrt{a^2-x^2}} dx = arcsin\frac{x}{a} + c$

19)$\int \frac{1}{a^2-x^2} dx = \frac{1}{2a} ln\left|\frac{a+x}{a-x}\right| + c$

20)$\int \frac{1}{\sqrt{a^2+x^2}} dx = ln|x + \sqrt{a^2 + x^2}| + c$

21)$\int \frac{1}{\sqrt{x^2-a^2}} dx = ln|x + \sqrt{x^2 - a^2}| + c$

22)$\int a^u\ du = \frac{a^u}{lna} + c$

23)$\int e^x\ dx = e^x + c$

24) $\int e^u du = e^u + c$

25) $\int e^{ax}\, dx = \frac{1}{a} e^{ax} + c$

26) $\int \sinh x\, dx = \cosh x + c$

27) $\int \cosh x\, dx = \sinh x + c$

28) $\int (1 - \tanh^2 x)\, dx = \int \frac{1}{\cosh^2 x}\, dx = \int \operatorname{sech}^2 x\, dx = \tanh x + c$

29) $\int (1 - \coth^2 x)\, dx = \int \frac{-1}{\sinh^2 x}\, dx = \int -\operatorname{csch}^2 x\, dx = \coth x + c$

30) $\int \operatorname{sech} x \tanh x\, dx = -\operatorname{sech} x + c$

31) $\int \operatorname{csch} x \coth x\, dx = -\operatorname{csch} x + c$

32) $\int \sin ax\, dx = \frac{-1}{a} \cos ax + c$

33) $\int \cos ax\, dx = \frac{1}{a} \sin ax + c$

34) $\int \sec^2 ax\, dx = \frac{1}{a} \tan ax + c$

35) $\int \csc^2 ax\, dx = \frac{-1}{a} \cot ax + c$

36) $\int \csc ax \cot ax\, dx = \frac{-1}{a} \csc ax + c$

37) $\int \sec ax \tan ax\, dx = \frac{1}{a} \sec ax + c$

38) $\int \sin^n x\, dx = -\frac{\sin^{n-1} x \cos x}{n} + \frac{n-1}{n} \int \sin^{n-2} x\, dx$

39) $\int \cos^n x\, dx = \frac{\cos^{n-1} x \sin x}{n} + \frac{n-1}{n} \int \cos^{n-2} x\, dx$

40) $\int \tan^n x\, dx = \frac{\tan^{n-1} x}{n-1} - \int \tan^{n-2} x\, dx$

41) $\int \cot^n x \, dx = -\dfrac{\cot^{n-1} x}{n-1} - \int \cot^{n-2} x \, dx$

42) $\int \sec^n x \, dx = \dfrac{\sec^{n-2} x \tan x}{n-1} + \dfrac{n-2}{n-1} \int \sec^{n-2} x \, dx$

43) $\int \csc^n x \, dx = -\dfrac{\csc^{n-2} x \cot x}{n-1} + \dfrac{n-2}{n-1} \int \csc^{n-2} x \, dx$

Note: In the above formulas a is a fixed value

Note 2: In the above formulas u is a function of x.

Integral

Integration is the opposite of differentiation.

Consider the following function:

$Y = x^2 + 2$

If we are asked to differentiate the above function the result will be 2x. Now what if we are asked to find the original function the derivative of which is 2x? In order to answer this question we need to integrate. Always remember the following formula which is the most important integration formula.

$$\int x^n dx = \frac{x^{n+1}}{n+1} + c$$

Also the term dx in front of the integration function means the differential of x.

Note: in multiplying fractions the numerator is multiplied by the numerator and the denominator is multiplied by the denominator. Also in multiplication and division the signs affect each other:

$- \times - = +, + \times + = +$

$- \times + = -, + \times - = -$

Example) $\frac{2}{3} \times \frac{5}{2} = \frac{2 \times 5}{3 \times 2} = \frac{10}{6}$

Example 2) $\frac{-1}{2} \times \frac{3}{8} = \frac{-1 \times 3}{2 \times 8} = \frac{-3}{16} = -\frac{3}{16}$

$Note: \dfrac{-a}{b} = \dfrac{a}{-b} = -\dfrac{a}{b}$

Example $\frac{-5}{3} = \frac{5}{-3} = -\frac{5}{3}$

Also in dividing two fractions the distant ones are multiplied by with other and the close one are multiplied by each other:

Example) $\frac{\frac{2}{7}}{\frac{3}{5}} = \frac{2\times5}{7\times3} = \frac{10}{21}$

Example 2) $\frac{\frac{x}{2}}{\frac{x^2}{2}} = \frac{x\times2}{2\times x^2} = \frac{2x}{2x^2} = \frac{1}{x}$

Note: in multiplying integer exponents with the same bases one of the bases is written and the powers are added. Also in dividing integer exponents with the same bases one of the bases is written and the powers are deducted.

$$a^n \times a^m = a^{n+m}$$

$$\frac{a^n}{a^m} = a^{n-m}$$

Example) $x^2 \times x^3 = x^{2+3} = x^5$

Example 2) $\frac{x^5}{x^3} = x^{5-3} = x^2$

Let's get back to integration:

$\int x^3 dx = \frac{x^{3+1}}{3+1} + c = \frac{x^4}{4} + c = \frac{1}{4}x^4 + c$

$\int x^7 dx = \frac{x^{7+1}}{7+1} + c = \frac{x^8}{8} + c = \frac{1}{8}x^8 + c$

$$Note: a^{-1} = \frac{1}{a}$$

Example) $2^{-1} = \frac{1}{2}$

Example 2) $x^{-1} = \frac{1}{x}$

$$\int x^{-1}dx = \int \frac{1}{x}dx = ln|x| + c$$

$$\int \frac{1}{x^3}dx = \int x^{-3}dx = \frac{x^{-3+1}}{-3+1} + c = \frac{x^{-2}}{-2} + c = -\frac{1}{2x^2} + c$$

$$\boxed{Note: \sqrt[m]{a^n} = a^{\frac{n}{m}}}$$

Example) $\sqrt[3]{x^2} = x^{\frac{2}{3}}$

Example 2) $\sqrt[4]{x^7} = x^{\frac{7}{4}}$

If a radical did not have an order, the order equals 2:

Example 3) $\sqrt{x^5} = \sqrt[2]{x^5} = x^{\frac{5}{2}}$

$$\int \sqrt{x}\,dx = \int x^{\frac{1}{2}}dx = \frac{x^{\frac{1}{2}+1}}{\frac{1}{2}+1} + c = \frac{x^{\frac{3}{2}}}{\frac{3}{2}} + c = \frac{x^{\frac{3}{2}}}{\frac{3}{2}} + c$$

$$= \frac{2x^{\frac{3}{2}}}{3} + c = \frac{2}{3}x^{\frac{3}{2}} + c = \frac{2}{3}\sqrt{x^3} + c$$

One might ask how I could calculate $1+\frac{1}{2}$; in response we use least common divisor.

Note in terms of least common divisor: consider $\frac{3}{6}$ and$\frac{4}{3}$: If we want to add up these numbers, we consider the smallest number that is divisible by both fraction denominator as a common denominator. Obviously we choose 6 the numbers 3 and 6 as the common denominator:

$$\frac{4}{3} + \frac{3}{6} = \frac{}{6}$$

In the second phase we divide 6 by the denominator of first fraction and multiply the first fraction numerator by the result:

$$\frac{6}{3} = 2 \times 4 = 8$$

Then do the same thing with the second fraction and the resulting numbers are added together:

$$\frac{6}{6} = 1 \times 3 = 3$$

The final answer: $\frac{3+8}{6} = \frac{11}{6}$

$\cdot \int \sqrt[5]{x^3}\,dx = \int x^{\frac{3}{5}}\,dx = \frac{x^{\frac{8}{5}}}{\frac{8}{5}} + c = \frac{x^{\frac{8}{5}}}{\frac{1}{\frac{8}{5}}} + c = \frac{5}{8}x^{\frac{8}{5}} + c = \frac{5}{8}\sqrt[5]{x^8} + c$

Note: In integration the constant is always transferred to the back of integral and it is multiplied by the result of integration.

$\cdot \int 2x^3\,dx = 2\int x^3\,dx = 2 \times \frac{x^4}{4} + c = \frac{2x^4}{4} + c = \frac{x^4}{2} + c = \frac{1}{2}x^4 + c$

$\cdot \int 5x^4\,dx = 5\int x^4\,dx = 5 \times \frac{x^{4+1}}{4+1} + c = 5 \times \frac{x^5}{5} + c = \frac{5x^5}{5} + c = x^5 + c$

Note: If we are faced with a polynomial function for integration, we calculate the integral of each term separately and then add or reduce them.

$. \int x^3 + 2x^2 + 1 \, dx = \int x^3 \, dx + \int 2x^2 \, dx + \int 1 \, dx =$

$\int x^3 \, dx + 2 \int x^2 \, dx + 1 \int dx = \frac{1}{4}x^4 + 2 \times \frac{x^3}{3} + x + c =$

$\frac{1}{4}x^4 + \frac{2}{3}x^3 + x + c$

Perhaps you might ask how the resulting integral of 1 dx was x? As mentioned before integration is the opposite of differentiation: Consider the following function:

y=x→y´=1

For integration we just need to go back: $\int 1 \, dx = x + c$

$. \int 2\sqrt{x} + \frac{4}{x^2} + \sqrt[3]{x^8} - 7 \, dx = 2 \int x^{\frac{1}{2}} \, dx + 4 \int \frac{1}{x^2} \, dx +$

$\int x^{\frac{8}{3}} \, dx - 7 \int dx = 2 \int x^{\frac{1}{2}} \, dx + 4 \int x^{-2} \, dx + \int x^{\frac{8}{3}} \, dx -$

$7 \int dx = 2 \times \frac{x^{\frac{3}{2}}}{\frac{3}{2}} + 4 \times \frac{x^{-1}}{-1} + \frac{x^{\frac{11}{3}}}{\frac{11}{3}} - 7 \times x + c = \frac{4}{3}\sqrt{x^3} -$

$\frac{4}{x} + \frac{3}{11}\sqrt[3]{x^{11}} - 7x + c$

One might ask, how the result of+ $4 \times \frac{x^{-1}}{-1}$ **is a negative value? Let's find the result of** $\frac{x^{-1}}{-1}$ **first:**

$$\frac{x^{-1}}{-1} = -\frac{1}{x}$$

In the second step+4 is multiplied by the above term: as mentioned before in multiplying fractions the denominators, numerator and signs affect the corresponding values:

$$+4 \times -\frac{1}{x} = -\frac{4}{x}$$

$$\int 3\sqrt[4]{x^3} + \frac{7}{x^5} + \frac{1}{6\sqrt{x}} dx$$

$$= 3\int x^{\frac{3}{4}}dx + 7\int \frac{1}{x^5}dx + \int \frac{1}{6x^{\frac{1}{2}}}dx = 3 \times \frac{x^{\frac{7}{4}}}{\frac{7}{4}}$$

$$+ 7 \times \frac{x^{-4}}{-4} + \frac{1}{6} \times \frac{x^{\frac{1}{2}}}{\frac{1}{2}} + c$$

$$= \frac{12}{7}x^{\frac{7}{4}} - \frac{7}{4x^4} + \frac{2}{6}\sqrt{x} + c$$

$$= \frac{12}{7}\sqrt[4]{x^7} - \frac{7}{4x^4} + \frac{1}{3}\sqrt{x} + c$$

Always keep in mind the following important factorizations:

Full Square 1)$(a + b)^2 = a^2 + 2ab + b^2$

Full Square 2)$(a - b)^2 = a^2 - 2ab + b^2$

Sum of the squares of two terms: 3) $a^2 + b^2 = (a + b)^2 - 2ab$

Difference/ sum of two square 4) (a-b)(a+b)=$a^2 - b^2$

One common term 5)(x+a)(x+b)=$x^2 + (a + b)x + ab$

Sum of three terms' square 6) $(a + b + c)^2 = a^2 + b^2 + c^2 + 2ab + 2ac + 2b$

Full cube 7)$(a + b)^3 = a^3 + 3a^2b + 3ab^2 + b^3$

Full cube 8)$(a - b)^3 = a^3 - 3a^2b + 3ab^2 - b^3$

Sum of the cubes of two terms: 9)$a^3 + b^3 = (a + b)(a^2 - ab + b^2)$

Difference of the cubes of two terms: 10)$a^3 - b^3 = (a - b)(a^2 + ab + b^2)$

$$\int \frac{(x + 2)^2}{\sqrt[3]{x^4}} dx = \int \frac{(x + 2)^2}{x^{\frac{4}{3}}} dx =$$

The numerator is a full square:

$$\int \frac{(x^2 + 4x + 4)}{x^{\frac{4}{3}}} dx =$$

At this stage $x^{\frac{4}{3}}$ is transferred into the numerator and multiplied by each term. Obviously the exponent sign will change:

$$\int (x^2 + 4x + 4)x^{-\frac{4}{3}}dx = \int (x^{\frac{2}{3}} + 4x^{-\frac{1}{3}} + 4x^{-\frac{4}{3}})dx$$

$$= \int x^{\frac{2}{3}} dx + 4\int x^{-\frac{1}{3}} dx + 4\int x^{-\frac{4}{3}} dx$$

$$= \frac{x^{\frac{5}{3}}}{\frac{5}{3}} + 4 \times \frac{x^{\frac{2}{3}}}{\frac{2}{3}} + 4 \times \frac{x^{-\frac{1}{3}}}{-\frac{1}{3}} + c$$

$$= \frac{x^{\frac{5}{3}}}{\frac{5}{3}} + \frac{12}{2}x^{\frac{2}{3}} - 12x^{-\frac{1}{3}} + c$$

$$= \frac{3}{5}\sqrt[3]{x^5} + 6\sqrt[3]{x^2} - \frac{12}{\sqrt[3]{x}} + c$$

$.\int \frac{x^2 - 9}{(x - 3)} dx =$

The numerator is a Sum of the squares of two terms:

$$\int \frac{(x-3)(x+3)}{(x-3)} dx =$$

(x-3) in numerator and denominator are crossed out then:

$$\int x + 3\, dx = \frac{x^4}{4} + 3x + c = \frac{1}{4}x^4 + 3x + c$$

Integration by changing the variable:

In most cases integration is not so easy and there are disturbing terms that make integration more complex. To respond to these questions: we consider a part of front integral function as u and then integrate it to obtain the result based on u. At the end of the problem we use the value of u in terms of x and calculate the final result. Let's discuss this with some examples: always keep in mind the following important formula:

$$\boxed{\int u^n du = \frac{u^{n+1}}{n+1} + c}$$

Note: u is a function of x.

$$.\int (x+3)^{13} dx =$$

Step 1: consider $x + 3$ as u and differentiate it:

X+3=u→1dx=du→dx=du

Step Two: Solve the integral in terms of u:

$$.\int u^{13} du = \frac{u^{14}}{14} + c = \frac{1}{14}u^{14} + c$$

Step three: Now we replace the value of u in terms of x:

U=x+3→$\frac{1}{14}(x+3)^{14}+c$

$\int 8x\sqrt{4x^2+3}\,dx=$

$4x^2+3=u$→8x dx=du

$\int \sqrt{u}\,du = \int u^{\frac{1}{2}}du = \frac{u^{\frac{3}{2}}}{\frac{3}{2}}+c = \frac{2}{3}\sqrt{u^3}+c$

Now we put the value of x instead of u:

$=\frac{2}{3}\sqrt{(4x^2+3)^3}+c$

$$\boxed{Note: \int \frac{du}{u} = \ln|u|+c}$$

$\int \frac{10x}{5x^2+7}\,dx=$

We use the method of change of variables:

Step 1: consider $5x^2+7$ equal to u and differentiate it:

$5x^2+7$=u→10x dx=du

$\int \frac{du}{u}=\ln|u|+c$

Step 2: Now we replace the value of u in terms of x:

U=$5x^2+7$→$\ln|5x^2+7|+c$

$\int \frac{xdx}{\sqrt{2x^2+5}}=$

We use the method of change of variables:

Step 1: consider $2x^2+5$ equal to u and differentiate it:

$2x^2 + 5 = u \rightarrow 4x\,dx = du$

We need xdx thus we divide both side by 4:

$X\,dx = \frac{1}{4}du$

Step 2: We solve the integral based on u:

$$\int \frac{\frac{1}{4}du}{\sqrt{u}} = \frac{1}{4}\int \frac{du}{u^{\frac{1}{2}}} = \frac{1}{4}\int u^{-\frac{1}{2}}du = \frac{1}{4} \times \frac{u^{\frac{1}{2}}}{\frac{1}{2}} + c = \frac{2}{4}u^{\frac{1}{2}} + c$$

$$= \frac{1}{2}\sqrt{u} + c$$

Step 3: Now we replace the value of u in terms of x:

$2x^2 + 5 = u \rightarrow \frac{1}{2}\sqrt{2x^2 + 5} + c$

You might ask which term should be considered to be equal to u. In response, I must say that the term equal to u is the one that its differentiation would provide the other side of the problem based on u and also in case of the above problem $2x^2 + 5$ is under the radical it should be equal to u.

$$\int \frac{x\,dx}{\sqrt{1 - 4x^2}} =$$

We use the method of change of variables:

Step 1: consider $1 - 4x^2$ equal to u and differentiate it:

$1 - 4x^2 = u \rightarrow -8x\,dx = du$

We need xdx thus we divide both side by -8.

$X\,dx = -\frac{1}{8}du$

Step 2: We solve the integral based on u:

$$.\int \frac{-\frac{1}{8}}{\sqrt{u}}\,du = -\frac{1}{8}\int \frac{du}{u^{\frac{1}{2}}} = -\frac{1}{8}\int u^{-\frac{1}{2}}\,du = -\frac{1}{8}\times\frac{u^{\frac{1}{2}}}{\frac{1}{2}}+c = -\frac{2}{8}u^{\frac{1}{2}}+$$

$$c = -\frac{1}{4}\sqrt{u}+c$$

Step 3: Now we replace the value of u in terms of x:

$$1-4x^2 = u \longrightarrow =-\frac{1}{4}\sqrt{1-4x^2}+c$$

Integration of trigonometric functions

Noting the discussion of differentiation:

Y=sinx→y´=cosx

Y=cosx→y´=-sinx

And integration is the opposite of differentiation:

$.\int sinx\,dx = -cosx + c$

$.\int cosx\,dx = sinx + c$

Also in trigonometry we remember:

Tanx=$\frac{sinx}{cosx}$, cotx=$\frac{cosx}{sinx}$

$.\int tanx\,dx = \int \frac{sinx}{cosx}\,dx =$

We use the method of change of variables:

Step 1: consider cosx equal to u and differentiate it:

Cosx=u→-sinx dx=du

We need sinx dx thus we multiply both side by -1.

Sinx dx=-du

$$=\int \frac{-du}{u} = -\int \frac{du}{u} = -\ln|u| + c$$

Step 3: Now we replace the value of u in terms of x:

Cosx=u→=-ln|cosx| + c

$. \int cotx \, dx = \int \frac{cosx}{sinx} \, dx =$

Step 1: consider sinx equal to u and differentiate it:

Sinx=u→cosx dx=du

Step 2: We solve the integral based on u:

$$\int \frac{du}{u} = \ln|u| + c$$

Step 3: Now we replace the value of u in terms of x:

=ln|sinx| + c

Prove) $\int sinax \, dx = \frac{-1}{a} cosax + c$

Solution: We use the method of change of variables:

Step 1: consider ax equal to u and differentiate it:

$ax=u \rightarrow a \, dx = du$

We need dx thus we divide both side by a.

$dx=\frac{1}{a} \, du$

$=\int sinu \times \frac{1}{a} \, du = \frac{1}{a}\int sinu \, du = \frac{1}{a} \times -cosu + c =$
$-\frac{1}{a} cosu + c$

Now we replace the value of u in terms of x:

$ax=u \rightarrow =-\frac{1}{a} cosax + c$

Note: You may wonder how we found that $\frac{1}{a}$ is a constant number and should be transferred behind the integral? In response it should be noted that we consider the letter in front of d: which is u in this problem. So anything else that is not based on u is a constant number and it should be transferred behind the integral.

Prove) $\int \cos ax\, dx = \frac{1}{a}\sin ax + c$

Solution: We use the method of change of variables:

Step 1: consider ax equal to u and differentiate it:

$ax=u\rightarrow a\,dx=du$

We need dx thus we divide both side by a.

$dx=\frac{1}{a}\,du$

Step 2: We solve the integral based on u:

$=\int \cos u\times\frac{1}{a}\,du=\frac{1}{a}\int \cos u\,du=\frac{1}{a}\times\sin u + c$

Step 3: Now we replace the value of u in terms of x:

$ax=u\rightarrow=\frac{1}{a}\sin ax + c$

Example ١) $\int \sin 4x\, dx = -\frac{1}{4}\cos 4x + c$

Example ٢) $\int \sin\frac{x}{3}dx=\int \frac{1}{3}\times dx=\frac{-1}{\frac{1}{3}}\cos\frac{1}{3}\times +c=\frac{\frac{-1}{1}}{\frac{1}{3}}\cos\frac{x}{3}+$ $c = -3\cos\frac{x}{3}+c$

Example ٣) $\int \cos 2x\, dx = \frac{1}{2}\sin 2x + c$

Example $\digamma$)$\int \cos 7x\, dx = \frac{1}{7}\sin 7x + c$

$$\int \frac{\sin x\, dx}{\sqrt{1 + \cos x}}$$

We use the method of change of variables:

Step 1: consider 1+cosx equal to u and differentiate it:

1+cosx=u→-sinx dx=du

We need sinx dx thus we multiply both side by -1.

Sinx dx=- du

Step 2: We solve the integral based on u:

$$= \int \frac{-du}{\sqrt{u}} = -\int \frac{du}{u^{\frac{1}{2}}} = -\int u^{-\frac{1}{2}}\, du = -\frac{u^{\frac{1}{2}}}{\frac{1}{2}} + c = -2\sqrt{u} + c$$

Step 3: Now we replace the value of u in terms of x:

1+cosx=u→=-2$\sqrt{1 + \cos x}$+c

.$\int \cos^{10}x \sin x\, dx =$

We use the method of change of variables:

Step 1: consider cosx equal to u and differentiate it:

Cosx=u→-sin x dx=du

We need sinx dx thus we multiply both side by -1.

Sinx dx=-du

Step 2: We solve the integral based on u:

$$= \int u^{10} \times -du = -\int u^{10} du = -\frac{u^{11}}{11} + c$$

Step 3: Now we replace the value of u in terms of x:

$$\text{Cosx=u} \rightarrow = -\frac{\cos^{11}x}{11} + c$$

$$.\int \frac{\sin x \, dx}{5+2\cos x} =$$

We use the method of change of variables:

Step 1: consider 5+2cosx equal to u and differentiate it:

Note: The derivation of the constant number (5) is zero. The derivation of the term 2cosx is the derivation of the two multiplied terms. Always remember the following formula:

$$\boxed{y = uv \rightarrow y' = u'v + v'u}$$

Y=2cosx→   u=2,cosx=v,0=u´,-sinx=v´

$$y' = 0 \times \cos x + (-\sin x) \times 2 = 0 + (-2\sin x) = -2\sin x$$

5+2cosx=u→-2sinx dx=du

We need sinx dx thus we divide both side by -2.

Sinx dx=$-\frac{1}{2}$ du

Step 2: We solve the integral based on u:

$$= \int \frac{-\frac{1}{2}}{u} \, du = -\frac{1}{2} \int \frac{du}{u} = -\frac{1}{2}\ln|u| + c$$

Step 3: Now we replace the value of u in terms of x:

5+2cosx=u→$= -\frac{1}{2}\ln|5 + 2\cos x|$+c

Reminder: Trigonometry

θ	0°	30°	45°	60°	90°
sin θ	0	$\frac{1}{2}$	$\frac{1}{\sqrt{2}}$	$\frac{\sqrt{3}}{2}$	1
cos θ	1	$\frac{\sqrt{3}}{2}$	$\frac{1}{\sqrt{2}}$	$\frac{1}{2}$	0
tan θ	0	$\frac{1}{\sqrt{3}}$	1	$\sqrt{3}$	∞
cot θ	∞	$\sqrt{3}$	1	$\frac{1}{\sqrt{3}}$	0
sec θ	1	$\frac{2}{\sqrt{3}}$	$\sqrt{2}$	2	∞
cosec θ	∞	2	$\sqrt{2}$	$\frac{2}{\sqrt{3}}$	1

Trigonometric ratio

Angle

Degree

Radian

Undefined

Consider the right triangle below:

a: opposite

b: adjacent

h: hypotenuse

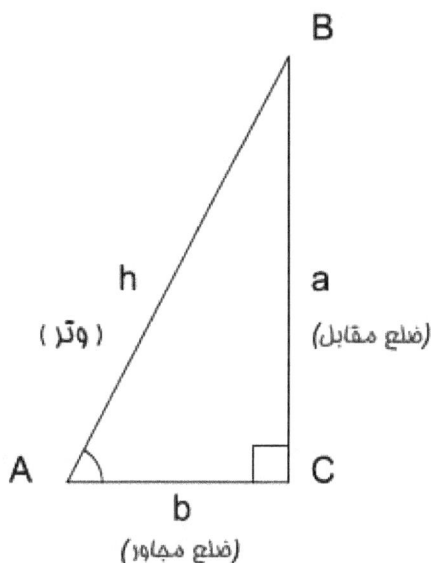

B

h
(وتر)

a
(ضلع مقابل)

A

b
(ضلع مجاور)

C

Hypotenuse is the opposite side of 90 degrees angle in right triangle

Pythagorean Theorem: states that the size of the hypotenuse powered by 2 equals with the sum of the squares of two other sides

$$h^2 = a^2 + b^2$$

$Sin(A) = \dfrac{opposite\ side}{hypotenuse} = \dfrac{a}{h}$

$Cos(A) = \dfrac{adjacent\ side}{hypotenuse} = \dfrac{b}{h}$

$Tan(A) = \dfrac{opposite\ side}{adjacent\ side} = \dfrac{a}{b}$

$Cot(A) = \dfrac{adjacent\ side}{opposite\ side} = \dfrac{b}{a}$

Arc: Suppose someone asks you where is the capital of france? Obviously you will quickly answer paris. Now someone may ask the same question in another form: Which country has paris as its capital? Again, you answer quickly france.

Suppose someone asks you Sin $(\frac{\pi}{4})$=? You will answer$(\frac{\sqrt{2}}{2})$. Now what if he asks arcsin $(\frac{\sqrt{2}}{2})$=?

arcsin$(\frac{\sqrt{2}}{2})$ means that which angle has the sinus of $\frac{\sqrt{2}}{2}$? The answer is $\frac{\pi}{4}$.

$$Note: arcsinx = sin^{-1}x$$

Example) arcsin$(\frac{1}{2}) = \frac{\pi}{6}$

Example 2) $\arccos(0)=\frac{\pi}{2}$

$.\int \frac{sinx}{cos^3 x}dx=$

We use the method of change of variables:

Step 1: consider cosx equal to u and differentiate it:

Cosx=u→-sinx dx=du

We need sinx dx thus we multiply both side by -1.

Sinx dx=-du

$=\int \frac{-du}{u^3}=$

Step 2: We solve the integral based on u:

$=- \int u^{-3}du=- \times \frac{u^{-2}}{-2}+c=- \times -\frac{1}{2u^2}+c=\frac{1}{2u^2}+c$

Step 3: Now we replace the value of u in terms of x:

$Cosx=u→=\frac{1}{2cos^2 x}+c$

Integration of odd exponents of sin and cos

Before the issue of the integration of odd exponents of sin and cos it is necessary to point out some issues:

$$\boxed{Note: (a^n)^m = a^{n \times m}}$$

Example) $(x^2)^3=x^{2\times3}$

Example 2) $(sin^2 x)^3=sin^{2\times3}x = sin^6 x$

Also we remember from trigonometry topics:

$sin^2 x+cos^2 x=1 \overset{so}{→} sin^2 x = 1 - cos^2 x \ and \ cos^2 x = 1 - sin^2 x$

$Secx=\frac{1}{\cos x}\rightarrow sec\ x\times\cos x=1$

Prove) $1+tan^2x=sec^2x\rightarrow tan^2x=\frac{sin^2x}{cos^2x}$

$1+\frac{sin^2x}{cos^2x}=$

We replace 1 with $\frac{cos^2x}{cos^2x}$:

$\frac{cos^2x}{cos^2x}+\frac{sin^2x}{cos^2x}\_\frac{sin^2x+cos^2x}{cos^2x}\_\frac{1}{cos^2x}=sec^2x$

Prove)$sec^2x-1=tan^2x$

We replace 1 with $\frac{cos^2x}{cos^2x}$ and replace sec^2x with $\frac{1}{cos^2x}$:

$\frac{cos^2x}{cos^2x}+\frac{sin^2x}{cos^2x}\_\frac{sin^2x+cos^2x}{cos^2x}\_\frac{1}{cos^2x}=sec^2x$

$\frac{1}{cos^2x}-\frac{cos^2x}{cos^2x}=\frac{1-cos^2x}{cos^2x}\_\frac{sin^2x}{cos^2x}=tan^2x$

Integration of odd exponents of sin and cos:

$.\int sin^3\ x\ dx=\int sin^2\ xsinx\ dx=$

Note)$a^n\times a^m=a^{n+m}\overset{so}{\rightarrow}a^{n+m}=a^n\times a^m$

Example)$sin^2x\times sinx=sin^3x\overset{so}{\rightarrow}sin^3\ x=sin^2x\times sinx$

We replace sin^2x with$1-cos^2x$:

$\int(1-cos^2\ x)sinx\ dx=$

We use the method of change of variables:

Step 1: consider cosx equal to u and differentiate it:

Cosx=u→-sinx dx=du

We need sinx dx thus we multiply both side by -1.

Sinx dx=-du

Step 2: We solve the integral based on u:

$.\int (1 - u^2) \times -du = -\int 1 - u^2\, du = -\int 1 du + \int u^2\, du =$
$-u + \frac{1}{3} u^3 + c$

Step 3: Now we replace the value of u in terms of x:

$.\cos x = u \rightarrow -\cos x + \frac{1}{3} \cos^3 x + c$

$.\int \cos^3 x\, dx = \int \cos^2 x \cos x\, dx =$

We replace $\cos^2 x$ with $1 - \sin^2 x$:

$= \int (1 - \sin^2 x) \cos x\, dx =$

We use the method of change of variables:

Step 1: consider sinx equal to u and differentiate it:

Sinx=u→cosx dx=du

Step 2: We solve the integral based on u:

$$\int (1 - u^2) du = \int 1\, du - \int u^2\, du = u - \frac{u^3}{3} + c$$
$$= u - \frac{1}{3} u^3 + c$$

Step 3: Now we replace the value of u in terms of x:

$\text{Sinx} = u \rightarrow \sin x - \frac{1}{3} \sin^3 x + c$

$.\int \sin^5 x\, dx = \int \sin^4 x \sin x\, dx = \int (\sin^2 x)^2 \sin x\, dx =$
$\int (1 - \cos^2 x)^2 \sin x\, dx =$

$1 - \cos^2 x$ is a full square thus:

$.\int (1 - 2\cos^2 x + \cos^4 x) \sin x\, dx =$

31

We use the method of change of variables:

Step 1: consider cosx equal to u and differentiate it:

Cosx=u→-sinx dx=du

We need sinx dx thus we multiply both side by -1.

Sinx dx=-du

$= \int (1 - 2u^2 + u^4) \times -du =$

Step 2: We solve the integral based on u:

$= -\int 1 - 2u^2 + u^4 \, du = -\int 1 \, du + 2 \int u^2 \, du - \int u^4 \, du =$

$-u + 2 \times \frac{u^3}{3} - \frac{u^5}{5} + c = -u + \frac{2}{3}u^3 - \frac{1}{5}u^5 + c$

Now we replace the value of u in terms of x:

Cosx=u→$= -\cos x + \frac{2}{3}\cos^3 x - \frac{1}{5}\cos^5 x + c$

$. \int \cos^5 x \, dx = \int \cos^4 x \cos x \, dx = \int (\cos^2 x)^2 \cos x \, dx =$
$\int (1 - \sin^2 x)^2 \cos x \, dx = \int (1 - 2\sin^2 x + \sin^4 x)\cos x \, dx =$

We use the method of change of variables:

Step 1: consider sinx equal to u and differentiate it:

Sinx=u→cosx dx=du

$= \int 1 \, du - 2 \int u^2 du + \int u^4 \, du =$

Step 2: We solve the integral based on u:

$= u - 2 \times \frac{u^3}{3} + \frac{u^5}{5} + c = u - \frac{2}{3}u^3 + \frac{1}{5}u^5 + c$

Step 3: Now we replace the value of u in terms of x:

Sinx=u→$= \sin x - \frac{2}{3}\sin^3 x + \frac{1}{5}\sin^5 x + c$

Integration of even exponents of sin and cos

Always remember the following trigonometric equations:

$$sin^2x = \frac{1 - \cos 2x}{2}$$

$$cos^2x = \frac{1 + \cos 2x}{2}$$

$.\int sin^2 x\, dx = \int \frac{1-\cos 2x}{2}\, dx =$

The common denominator is the number 2 therefore the fractions can be separated:

$= \int \frac{1}{2} - \frac{\cos 2x}{2}\, dx = \frac{1}{2}\int dx - \frac{1}{2}\int \cos 2x\, dx = \frac{1}{2}\,x - \frac{1}{2} \times$
$\frac{1}{2}\sin 2x + c = \frac{1}{2}\,x - \frac{1}{4}\sin 2x + c$

$.\int sin^2 x\, dx = \int (sin^2x)^2\, dx = \int (\frac{1-\cos 2x}{2})^2\, dx =$

$$Note: (\frac{a}{b})^n = \frac{a^n}{b^n}$$

$= \int \frac{(1-\cos 2x)^2}{2^2}\, dx = \int \frac{1-2\cos 2x+\cos^2 2x}{4}\, dx =$

The common denominator is the number 4 therefore the fractions can be separated:

$= \int \frac{1}{4} - \frac{2\cos 2x}{4} + \frac{\cos^2 2x}{4}\, dx = \frac{1}{4}\int dx - \frac{1}{2}\int \cos 2x +$
$\frac{1}{4}\int \cos^2 2x\, dx =$

$$Note: cos^2u = \frac{1 + \cos 2u}{2}$$

$$=\frac{1}{4}x-\frac{1}{2}\times\frac{1}{2}sin\,2x+\frac{1}{4}\int\frac{1+cos2(2x)}{2}=\frac{1}{4}x-\frac{1}{4}sin\,2x+\frac{1}{4}\int\frac{1}{2}+$$

$$\frac{1}{2}cos\,4x=$$

$$\frac{1}{4}\int\frac{1}{2}+\frac{1}{2}cos\,4x=\frac{1}{4}\int\frac{1}{2}\,dx+\frac{1}{4}\int\frac{1}{2}cos\,4x\,dx=\frac{1}{8}\int dx+$$

$$\frac{1}{8}\int cos\,4x\,dx=\frac{1}{8}x+\frac{1}{8}\times\frac{1}{4}sin\,4x+c=\frac{1}{8}x+\frac{1}{32}sin\,4x+c$$

The final answer: $=\frac{1}{4}x-\frac{1}{4}sin\,2x+\frac{1}{8}x+\frac{1}{32}sin\,4x+c=$

$$\frac{3}{8}x-\frac{1}{4}sin\,2x+\frac{1}{32}sin\,4x+c$$

Integration of sin and cos multiplication functions

To answer these questions always remember multiplication to sum conversion formulas

1)sin a cos b$=\frac{1}{2}[sin(a+b)+sin(a-b)]$

Example)$\int sin\,5x\,cos\,3x\,dx=$

$.sin\,5x\,cos\,3x=\frac{1}{2}[sin(5x+3x)+sin(5x-$

$3x)]=\frac{1}{2}[sin\,8x+sin\,2x]=$

$.\int\frac{1}{2}[sin\,8x+sin\,2x]dx=\frac{1}{2}\int sin\,8x\,dx+\frac{1}{2}\int sin\,2x\,dx=$

$\frac{1}{2}\times-\frac{1}{8}cos\,8x+\frac{1}{2}\times-\frac{1}{2}cos\,2x+c=-\frac{1}{16}cos\,8x-$

$\frac{1}{4}cos\,2x+c$

2)sinasinb$=-\frac{1}{2}[cos(a+b)-cos(a-b)]$

$.\int sin\,7x\,sin\,4x\,dx=$

$$sin\,7x\,sin\,4x=-\frac{1}{2}[cos(7x+4x)-cos(7x-4x]$$

$$=-\frac{1}{2}[cos\,11x-cos3x]$$

$. = \int -\frac{1}{2}[\cos 11x - \cos 3x]dx = -\frac{1}{2}\int \cos 11x \, dx +$

$\frac{1}{2}\int \cos 3x \, dx = -\frac{1}{2} \times \frac{1}{11}\sin 11x + \frac{1}{2} \times \frac{1}{3}\sin 3x + c =$

$-\frac{1}{22}\sin 11x + \frac{1}{6}\sin 3x + c$

3)cosacosb$=\frac{1}{2}[\cos(a+b) + \cos(a-b)]$

Example)$\int \cos 8x \cos 5x \, dx =$

$$\cos 8x \cos 5x = \frac{1}{2}[\cos(8x + 5x) + \cos(8x - 5x)]$$

$$= \frac{1}{2}[\cos 13x + \square os \, 3x]$$

$$= \int \frac{1}{2}[\cos 13x + \cos 3x]$$

$$= \frac{1}{2}\int \cos 13x \, dx + \frac{1}{2}\int \cos 3x \, dx = \frac{1}{2}$$

$$\times \frac{1}{13}\sin 13x + \frac{1}{2}$$

$$\times \frac{1}{3}\sin 3x + c = \frac{1}{26}\sin 13x + \frac{1}{6}\sin 3x + c$$

Integration by converting trigonometric variables

1) Integrals including $\sqrt{a^2 - x^2}$: If we factorize a^2 from the term under the radical the term would be like $a^2(1 - \frac{x^2}{a^2})$ which reminds us about 1-$sin^2\theta$: So we act as follows:

$. \int \sqrt{a^2 - x^2} \, dx = \int \sqrt{a^2(1 - \frac{x^2}{a^2})} \, dx$

Here $1 - \frac{x^2}{a^2}$ is considered to be 1-$sin^2\theta$:

$1 - \frac{x^2}{a^2} = 1\text{-}sin^2 \, \theta \rightarrow \frac{x^2}{a^2} = sin^2 \, \theta$

At this point we obtain the square root of both sides:

$$\frac{x}{a} = \sin\theta \rightarrow x = a\sin\theta$$

As you see for integrating $\sqrt{a^2 - x^2}$ we use $x = a\sin\theta$ variable:

Example: $\int \frac{1}{\sqrt{3-2x^2}}\, dx = -\frac{\pi}{2} \ll \theta \ll \frac{\pi}{2}$

We factorize 3 under the radical:

$$.\int \frac{1}{3(1-\frac{2x^2}{3})}\, dx =$$

We consider $1 - \frac{2x^2}{3}$ in the denominator equal to 1-$\sin^2\theta$:

$$1 - \frac{2x^2}{3} = 1-\sin^2\theta \rightarrow \frac{2}{3}x^2 = \sin^2\theta$$

Now we take the square root of both sides:

$$\frac{\sqrt{2}}{\sqrt{3}}x = \sin\theta$$

We divide $\sin\theta$ by 1 and cross multiply:

$$\sqrt{2}\,x \times 1 = \sqrt{3} \times \sin\theta \rightarrow \sqrt{2}\,x = \sqrt{3}\sin\theta$$

At this stage to find θ and x divide the known side by the unknown one:

$$x = \frac{\sqrt{3}\sin\theta}{\sqrt{2}} = \frac{\sqrt{3}}{\sqrt{2}}\sin\theta$$

$$\theta = \frac{\sqrt{2}\,x}{\sqrt{3}\sin\theta} = \frac{\sqrt{2}}{\sqrt{3}}x \times \frac{1}{\sin} = arc\,\sin(\frac{\sqrt{2}}{\sqrt{3}}x)$$

$.dx = \frac{\sqrt{3}}{\sqrt{2}} \cos\theta . d\theta$

Now we replace the obtained values in the integral:

$.\int \frac{1}{3(1-\sin^2\theta)} \times \frac{\sqrt{3}}{\sqrt{2}} \cos\theta . d\theta$

$$\boxed{Note: \sqrt[n]{a \times b} = \sqrt[n]{a} \times \sqrt[n]{b}}$$

$.\sqrt{3(1-\sin^2\theta)} = \sqrt{3} \times \sqrt{(1-\sin^2\theta)}$

$= \int \frac{1}{\sqrt{3} \times \sqrt{(1-\sin^2\theta)}} \times \frac{\sqrt{3}}{\sqrt{2}} \cos\theta . d\theta = \int \frac{\sqrt{3}\cos\theta d\theta.}{\sqrt{2} \times \sqrt{3} \times \sqrt{(1-\sin^2\theta)}} =$

Note:$\sin^2\theta + \cos^2\theta = 1 \rightarrow \cos^2\theta = 1 - \sin^2\theta$

Note:$\sqrt{a^2} = |a| \rightarrow \sqrt{\cos^2\theta} = |\cos\theta|$

$= \int \frac{\sqrt{3}\cos\theta d\theta.}{\sqrt{2} \times \sqrt{3} \times \sqrt{\cos^2\theta}} =$

As you can see in numerator our range of choice is $-\frac{\pi}{2}$ to $\frac{\pi}{2}$ and you know that cos is always positive in this range, therefore while driving $\cos^2\theta$ from the radical there is no need to put the absolute sign.

$.\int \frac{\sqrt{3}\cos\theta d\theta.}{\sqrt{2} \times \sqrt{3} \times \cos\theta} =$

$\sqrt{3}\cos\theta$ in the numerator and denominator are crossed out

$\int \frac{d\theta}{\sqrt{2}} = \frac{1}{\sqrt{2}} \int d\theta = \frac{1}{\sqrt{2}} \theta + c$

Now we replace the value of θ in terms of x:

2) Integrals including $a^2 + x^2$: To answer these questions we factorize a^2: the resulting term is $a^2(1 +$

$\frac{x^2}{a^2}$): The term in the parentheses reminds us about $1+tan^2\theta$. Thus we consider it equal to $1+tan^2\theta$: consider the following example:

$$.\int \frac{1}{25+x^2} \, dx =$$

We factorize of 25 in the denominator:

$$=\int \frac{1}{25(1+\frac{x^2}{25})} \, dx =$$

Now we take the square root of both sides:

$$\frac{x}{5} = tan \, \theta$$

We divide $tan \, \theta$ by 1 and cross multiply:

$$\frac{x}{5} = \frac{tan \, \theta}{1} \rightarrow x \times 1 = 5 \times tan \, \theta \rightarrow x = 5tan\theta \rightarrow dx = 5(1+tan^2\theta)d \qquad \theta$$

Note: $1+tan^2\theta = \frac{1}{cos^2\theta} = sec^2 \, \theta \rightarrow dx = 5sec^2 \, \theta. d \, \theta$

$$\theta = \frac{x}{5tan} = \frac{x}{5} \times \frac{1}{tan} = arc \, tan(\frac{x}{5})$$

Now we replace the obtained values in the integral:

$$=\int \frac{1}{25(1+tan^2\theta)} \times 5sec^2 \, \theta. d \, \theta = \int \frac{1}{25(5sec^2 \, \theta)} \times 5sec^2 \, \theta. d \, \theta =$$

$$\int \frac{5sec^2 \, \theta. d \, \theta}{25(5sec^2 \, \theta)} = \frac{5}{25} \int \frac{sec^2 \, \theta}{sec^2 \, \theta} \, d \, \theta = \frac{1}{5} \int 1 \, d \, \theta = \frac{1}{5}\theta + c \rightarrow$$

$$\theta = arc \, tan \left(\frac{x}{5}\right) \rightarrow = \frac{1}{5} arc \, tan \left(\frac{x}{5}\right) + c$$

The quick answer:

$$.\int \frac{1}{a^2+x^2} \, dx = \frac{1}{a} arc \, tan \frac{x}{a} + c$$

$\cdot \int \frac{1}{25+x^2}\, dx = \int \frac{1}{5^2+x^2} = \qquad\qquad\qquad\qquad$ a=5

$= \frac{1}{5} \arctan \frac{x}{5} + c$

3) Integrals including $\sqrt{x^2 - a^2}$: To answer these questions we use the change of variable x=asex θ.

$\boxed{Note: \sec^2 \theta - 1 = \tan^2\theta}$

Prove: $\int e^{ax}\, dx = \frac{1}{a}e^{ax} + c$

We use the method of change of variables:

Step 1: consider ax equal to u and differentiate it:

ax=u→a dx=du→dx=$\frac{1}{a}$ du

Step 2: We solve the integral based on u:

$$\int e^u \times \frac{1}{a}\, du = \frac{1}{a}\int e^u\, du = \frac{1}{a} \times e^u + c = \frac{1}{a}e^u + c$$

Step 3: Now we replace the value of u in terms of x:

ax=u→$=\frac{1}{a}e^{ax} + c$

$\cdot \int e^x \sec^2 (e^x)dx =$

We use the method of change of variables:

Step 1: consider e^x equal to u and differentiate it:

$e^x = u \rightarrow e^x dx = du$

Step 2: We solve the integral based on u:

$=\int \sec^2 u\, du = \tan u + c$

Step 3: Now we replace the value of u in terms of x:

$$e^x = u \to = tan(e^x) + c$$

$$\int e^x \sqrt{5 + 7e^x}\, dx =$$

We use the method of change of variables:

Step 1: consider $5 + 7e^x$ equal to u and differentiate it:

$$5 + 7e^x = u \to 7e^x\, dx = du \to e^x\, dx = \frac{1}{7}\, du$$

Step 2: We solve the integral based on u:

$$= \int \sqrt{u} \times \frac{1}{7}\, du = \frac{1}{7} \int u^{\frac{1}{2}}\, du = \frac{1}{7} \times \frac{u^{\frac{3}{2}}}{\frac{3}{2}} + c = \frac{2}{21} u^{\frac{3}{2}} + c =$$

$$\frac{2}{21} \sqrt{u^3} + c$$

Step 3: Now we replace the value of u in terms of x:

$$5 + 7e^x = u \to = \frac{2}{21} \sqrt{(5 + 7e^x)^3} + c$$

$$\int \frac{e^{\sqrt{x+5}}}{\sqrt{x+5}}\, dx =$$

We use the method of change of variables:

Step 1: consider $\sqrt{x + 5}$ equal to u and differentiate it:

$$\boxed{y = u^n \to y' = nu'u^{n-1}}$$

$$\sqrt{x + 5} = u$$

$$y = \sqrt{x+5} \to y = (x+5)^{\frac{1}{2}} \to y' = \frac{1}{2} \times (1) \times (x+5)^{-\frac{1}{2}} =$$

$$\frac{1}{2(x+5)^{\frac{1}{2}}} = \frac{1}{2\sqrt{x+5}}$$

$$\sqrt{x + 5} = u \to \frac{1}{2\sqrt{x+5}}\, dx = du \to \frac{1}{\sqrt{x+5}}\, dx = 2\, du$$

Step 2: We solve the integral based on u:

$$= \int e^u \times 2du = 2 \int e^u \, du = 2 \times e^u + c = 2e^u + c$$

Step 3: Now we replace the value of u in terms of x:

$$\sqrt{x+5} = u \rightarrow = 2e^{\sqrt{x+5}} + c$$

$$. \int \frac{sinx}{cos^3x} \, dx =$$

We use the method of change of variables:

Step 1: consider cosx equal to u and differentiate it:

$$\cos x = u \rightarrow -\sin x \, dx = du \rightarrow \sin x \, dx = -du$$

Step 2: We solve the integral based on u:

$$= \int \frac{-du}{u^3} = -\int u^{-3} \, du = -\times \frac{u^{-2}}{-2} + c = -\times -\frac{1}{2u^2} + c = \frac{1}{2u^2} + c$$

Step 3: Now we replace the value of u in terms of x:

$$\text{Cos } x = u \rightarrow = \frac{1}{2cos^2x} + c$$

$$. \int sec^3 xtan \, x \, dx =$$

Note: $sec^3x = sec^2xsec \, x$

We use the method of change of variables:

Step 1: consider $sec \, x$ equal to u and differentiate it:

$$= \int sec^2 \, xsec \, x \, tan \, x \, dx =$$

$$sec \, x = u \rightarrow sec \, xtan \, x \, dx = du$$

Step 2: We solve the integral based on u:

$$= \int u^2 \, du = \frac{u^3}{3} + c = \frac{1}{3}u^3 + c$$

Step 3: Now we replace the value of u in terms of x:

Sec x=u $\longrightarrow = \frac{1}{3} sec^3 x + c$

Always remember the following formula:

$$\int a^u du = \frac{a^u}{lna} + c$$

Note: log: $log_a y = x \Leftrightarrow a^x = y$

The logarithm of a number is the exponent to which another fixed value, the base, must be raised to produce that number.

Example $log_2 4 = ? \longrightarrow$ (Read the log of 4 to base 2.)

Now we ask where the result is 4 what is the power of 2? Obviously $2^2 = 4$. Thus the log of 4 to base 2 equals 2.

$$log_2 4 = 2$$

Example 2) $log_4 4 = ?$

We ask where the result is 4 what is the power of 4? Obviously $4^1 = 4$.

$4^1 = 4 \longrightarrow log_4 4 = 1$

Conclusion: The logarithm of any number to base itself equals 1.

Example 3) $log_8 1 = ?$

We ask where the result is 1 what is the power of 8?

We know that the result of any number powered by 0 equals 1.

$$a^0 = 1$$

Conclusion: Any number powered by 0 has the result of 1. $8^0 = 1$

$8^0 = 1 \rightarrow log_8 1 = 0$

Conclusion: The logarithm of 1 to any base equals 0.

Also, always keep in mind the following rules:

1) $log_a x \times y = log_a x + log_a y$

2) $log_a \frac{x}{y} = log_a x - log_a y$

Ln: Ln of any number means the logarithm of that number to the base e (Napier's constant). For example ln2 means $log_e 2$ (read the log of 2 to base e). These types of functions (the functions to the base e) are called natural logarithmic functions. e equals 2.7.

Note: all laws about the logarithm are true about Ln.

.$\int 8^x dx =$

We use the method of change of variables:

Step 1: consider x equal to u and differentiate it:

X=u→dx=du

Step 2: We solve the integral based on u:

$$\int 8^u du = \frac{8^u}{ln8} + c$$

Step 3: Now we replace the value of u in terms of x:

X=u→$=\frac{8^x}{ln8} + c$

43

$. \int \frac{7^{\sqrt{x}}}{2\sqrt{x}} \, dx =$

We use the method of change of variables:

Step 1: consider $\sqrt{x}$ equal to u and differentiate it:

$$\sqrt{x} = u \rightarrow \frac{1}{2\sqrt{x}} \, dx = du$$

Step 2: We solve the integral based on u:

$$= \int 7^u \, du = \frac{7^u}{\ln 7} + c$$

Step 3: Now we replace the value of u in terms of x:

$$\sqrt{x} = u \rightarrow = \frac{7^{\sqrt{x}}}{\ln 7} + c$$

$. \int 5^{3x} \, dx =$

We use the method of change of variables:

Step 1: consider $3x$ equal to u and differentiate it:

$3x=u \rightarrow 3 \ dx=du \rightarrow dx = \frac{1}{3} \, du$

Step 2: We solve the integral based on u:

$$= \int 5^u \times \frac{1}{3} \, du = \frac{1}{3} \int 5^u \, du = \frac{1}{3} \times \frac{5^u}{\ln 5} + c = \frac{5^u}{3\ln 5} + c$$

Step 3: Now we replace the value of u in terms of x:

$3x=u \rightarrow = \frac{5^{3x}}{3\ln 5} + c$

Fractional integration method

There are Integrals that cannot be solved by the method of change of variables because: it is not possible to create the other side by differentiating one side. The general form of these problems is such that usually two terms of different family are multiplied by each other. For example multiplying a trigonometric function by a polynomial:

$\int xsinx \, dx$

To answer these questions, we use fractional method:

For solving integral by fractional method use the following formula:

$$\int udv = uv - \int vdu$$

It is obvious that u and v are two functions of x.

As the formula suggests in such problems a part of the equation is u and the other part is considered as dv. We differentiate u to obtain du and dv which is the differential of a term is integrated to obtain v and then replace the obtained values in the formula. Let's discuss this with some examples:

$\int x \sin x \, dx =$

Solution: consider x equal to u and differentiate it to obtain du. The sin x dx which is the differential of a term is considered as dv and integrate it to obtain v.

X=u→dx=du

Sin x dx=dv→-cos x=v

We replace the obtained values in $\int u\,dv = uv - \int v\,du$

$=x \times -\cos x - \int -\cos x \times dx = -x\cos x + \int \cos x\,dx$

$.\int \cos x\,dx = \sin x + c$

The final answer$)=-x\cos x + \sin x + c$

Q) How do we determine which term is considered as u?

Answer: The answer to that there is no specific law in this area, and in answering integral problems the most important factor is experience. But the following steps may help:

The priority of choosing u:

1) Trigonometric Reverse: arcsin.arc cos and...

2) Log Functions: log and ln

3) Algebra: polynomials $x^3 - 2x$ and ...

4) Exponential functions: $e^x - 2^x$ and ...

5) Trigonometric Functions: sinx-cos x-tan x and ...

$.\int x \cos 2x\,dx =$

We use fractional method:

Consider x equal to u and differentiate it to obtain du. The cos 2x dx is considered as dv and integrate it to obtain v. We replace the obtained values in $\int u\,dv = uv - \int v\,du$

X=u→dx=du

Cos 2x dx=dv→$\frac{1}{2}\sin 2x = v$

$\int udv = uv - \int vdu$

$$= x$$

$$\times \frac{1}{2}\sin 2x - \int \frac{1}{2}\sin 2x \times dx$$

$$= \frac{1}{2}x\sin 2x - \frac{1}{2}\int \sin 2x\, dx =$$

$.\int \sin 2x\, dx = -\frac{1}{2}\cos 2x + c$

The final answer: $\frac{1}{2}x\sin 2x + \frac{1}{4}\cos 2x + c$

$.\int x\, lnx\, dx =$

We use fractional method:

Consider lnx equal to u and differentiate it. The xdx is considered as dv and integrate it to obtain v. We replace the obtained values in $\int udv = uv - \int vdu$

Lnx=u→$\frac{1}{x}\, dx = du$

$x\, dx = dv \rightarrow \frac{1}{2}x^2 = v$

$$\int udv = uv - \int vdu = lnx \times \frac{1}{2}x^2 - \int \frac{1}{2}x^2 \times \frac{1}{x}\, dx$$

$$= \frac{1}{2}x^2 lnx - \frac{1}{2}\int \frac{x^2}{x}\, dx$$

$.\int \frac{x^2}{x}\, dx = \int x\, dx = \frac{1}{2}x^2 + c$

The final answer: $\frac{1}{2}x^2 lnx - \frac{1}{4}x^2 + c$

$. \int \cos^{-1} x \, dx =$

We use fractional method:

Consider $\cos^{-1}x$ equal to u and differentiate it. The dx is considered as dv and integrate it to obtain v. We replace the obtained values in $\int u\,dv = uv - \int v\,du$

$$\cos^{-1}x = u \rightarrow -\frac{1}{\sqrt{1-x^2}}dx = du$$

$$dx = dv \rightarrow x = v$$

$$\int u\,dv = uv - \int v\,du = \cos^{-1}x \times x - \int x \times -\frac{1}{\sqrt{1-x^2}}\,dx$$

$$= x\cos^{-1}x + \int \frac{x\,dx}{\sqrt{1-x^2}}$$

$$\int \frac{x\,dx}{\sqrt{1-x^2}} =$$

We use the method of change of variables:

Step 1: Consider $1 - x^2$ equal to u and differentiate it.

$$1 - x^2 = u \rightarrow -2x\,dx = du \rightarrow x\,dx = -\frac{1}{2}\,du$$

Step 2: We solve the integral based on u:

$$= \int \frac{-\frac{1}{2}}{\sqrt{u}}\,du = -\frac{1}{2}\int \frac{du}{u^{\frac{1}{2}}} = -\frac{1}{2}\int u^{-\frac{1}{2}}\,du = -\frac{1}{2} \times \frac{u^{\frac{1}{2}}}{\frac{1}{2}} + c =$$
$$-\frac{2}{2}\sqrt{u} + c = -\sqrt{u} + c$$

Step 3: Now we replace the value of u in terms of x:

$$1 - x^2 = u \rightarrow = -\sqrt{1-x^2} + c$$

The final answer: $x\cos^{-1}x - \sqrt{1-x^2} + c$

$.\int \tan^{-1}x\, dx =$

We use fractional method:

Consider $\tan^{-1}x$ equal to u and differentiate it. The dx is considered as dv and integrate it to obtain v. We replace the obtained values in the formula:

$$\tan^{-1}x = u \rightarrow \frac{1}{1+x^2}\, dx = du$$

$$dx = dv \rightarrow x = v$$

$\int u\,dv = uv - \int v\,du = \tan^{-1}x \times x - \int x \times \frac{1}{1+x^2}\, dx =$
$x\tan^{-1}x - \int \frac{x}{1+x^2}dx$

$.\int \frac{x}{1+x^2}\, dx =$

We use the method of change of variables:

Step 1: Consider $1+x^2$ equal to u and differentiate it.

$$1+x^2 = u \rightarrow 2x\, dx = du \rightarrow x\, dx = \frac{1}{2}\, du$$

Step 2: We solve the integral based on u:

$$\int \frac{\frac{1}{2}du}{u} = \frac{1}{2}\int \frac{du}{u} = \frac{1}{2}\ln|u| + c$$

Step 3: Now we replace the value of u in terms of x:

$$1+x^2 = u \rightarrow= \frac{1}{2}\ln|1+x^2| + c$$

The final answer: $x\tan^{-1}x - \frac{1}{2}\ln|1+x^2| + c$

$. \int lnx \, dx =$

We use fractional method:

Consider *lnx* equal to u and differentiate it. The dx is considered as dv and integrate it to obtain v. We replace the obtained values in the formula:

$$lnx = u \rightarrow \frac{1}{x} \, dx = du$$

$$dx = dv \rightarrow x = v$$

$$\int udv = uv - \int vdu = lnx \times x - \int x \times \frac{1}{x} \, dx$$

$$= x \, lnx - \int 1 \, dx = xlnx - x + c$$

The use of table in fractional integration: a quick way to respond to fractional integration problems. In this method the two parts are written separately and differentiate it until we obtain 0 and again integriate the other one until it is placed in front of 0. Then we cross the right side to the left and assign+ and − every other one. Let me make it clear with some examples:

$. \int x^3 \sin x \, dx =$

	x^3 and its derivatives	$sin x$ and its integrates
+	x^3 ↘	Sin x
-	$3x^2$ ↘	-cos x
+	$6x$ ↘	-sin x

-	6↘	Cos x
	0	Sin x

As you can see each of the terms in the left is multiplied by the term beneath it in the right side and they are assigned + and − signs:

The final answer: $-x^3 \cos x + 3x^2 \sin x + 6x \cos x - 6 \sin x + c$

	x^4 and its derivatives	e^x and its integrates
+	x^4 ↘	e^x
-	$4x^3$ ↘	e^x
+	$12x^2$ ↘	e^x
-	$24x$ ↘	e^x
+	24 ↘	e^x
	0	e^x

As you can see each of the terms in the left is multiplied by the term beneath it in the right side and they are assigned + and − signs:

The final answer: $x^4 e^x - 4x^3 e^x + 12x^2 e^x - 24xe^x + 24e^x + c$

Recursive Integrals: There are integrals that after a few integrations we get back to the original point. These are called recursive Integrals. In order to solve such problems we consider the problem as i:

.i= $\int e^x \sin x \, dx =$

We use fractional method:

$e^x = u \rightarrow e^x dx = du$

$sinx\ dx = dv \rightarrow -\cos x = v$

$$\int udv = uv - \int vdu$$

$$= e^x$$

$$\times -cosx - \int -\cos x \times e^x\ dx$$

$$= -e^x \cos x + \int e^x \cos x\ dx =$$

$.\int e^x \cos x\ dx =$

We use fractional method again:

$e^x = u \rightarrow e^x dx = du$

$\cos x\ dx = dv \rightarrow \sin x = v$

$$\int udv = uv - \int vdu$$

$$= e^x$$

$$\times \sin x - \int \sin x \times e^x dx = e^x \sin x$$

$$- \int e^x \sin x\ dx$$

As you can see, after double integration $\int e^x \sin x\ dx$ was obtained again thus this term is considered as equal to I and act as follows:

$$i = -e^x \cos x + e^x \sin x - i \rightarrow i + i = -e^x \cos x + e^x \sin x$$
$$\rightarrow 2i$$
$$= -e^x \cos x$$
$$+ e^x \sin x \rightarrow 2i = e^x \sin x - e^x \cos x \rightarrow i$$
$$= \frac{e^x \sin x - e^x \cos x}{2}$$

$$i = \int \sin(lnx)\, dx =$$

We use fractional method again:

Note: $y = \sin u \rightarrow y' = u' \cos u$

$$y = \sin(lnx) \rightarrow lnx = u \rightarrow \frac{1}{x} = u' \rightarrow y' = \frac{1}{x}\cos(lnx)$$

$$\sin(lnx) = u \rightarrow \frac{1}{x}\cos(lnx)\, dx = du$$

$$dx = dv \rightarrow x = v$$

$$\int u\, dv = uv - \int v\, du$$

$$= \sin(lnx) \times x - \int x \times \frac{1}{x}\cos(lnx)\, dx$$

$$= x\sin(lnx) - \int \cos(lnx)\, dx$$

$.\int \cos(lnx)\, dx =$

We use fractional method again:

$$\cos(lnx) = u \rightarrow -\frac{1}{x}\sin(lnx)\, dx = du$$

$$dx = dv \rightarrow x = v$$

$$\int u\,dv = uv - \int v\,du$$

$$= \cos(lnx) \times x - \int x \times -\frac{1}{x}\sin(lnx)\,dx =$$

$$x\cos(lnx) + \int \sin(lnx)\,dx =$$

As you can see, after double integration we are back to the original point so we consider $\int \sin(lnx)\,dx$ as equal to i:

$$i = xsin(lnx) - xcos(lnx) - i$$

$$2i = xsin(lnx) - xcos(lnx)$$

$$i = \frac{xsin(lnx) - xcos(lnx)}{2}$$

Note: expansion of fractions:

Let me discuss expansion of fractions in detail:

$$\frac{5x-10}{(x-4)(x+1)} =$$

Fist draw two fraction lines and put a + sign between them put a and b in first and second numerator also put x-4 and x+1 in the first and second denominator. Then we obtain the common denominator:

$$\frac{a}{x-4} + \frac{b}{x+1} = \frac{a(x+1) + b(x-4)}{(x-4)(x+1)}$$

At this point we multiply the letters a and b by the parentheses:

$$\frac{ax + a + bx - 4b}{(x - 4)(x + 1)} =$$

Here we factorize x in the nominator:

$$\frac{x(a+b)+a-4b}{(x-4)(x+1)} =$$

Now we compare the nominator of the obtained fraction and the main fraction, the factor 5 in the denominator of the original fraction is 5, thus:

a+b=5

Also the constant in the original fraction is -10, thus:

a-4b=-10

We put the values in the system and solve it:

$$\begin{cases} a + b = 5 \\ a - 4b = -10 \end{cases}$$

We multiply the first line by -1 to remove a in the first and second row:

$$\begin{cases} -a - b = -5 \\ a - 4b = -10 \end{cases} \rightarrow -5b = -15 \rightarrow b = \frac{-15}{-5} = 3$$

$$a + b = 5 \rightarrow a + 3 = 5 \rightarrow a = 5 - 3 = 2$$

So after expanding the fraction will be as follows:

$$\frac{2}{x - 4} + \frac{3}{x + 1}$$

-Sometimes we use identities: consider the following fraction:

$$\frac{x}{x^2 - 5x + 6} =$$

The denominator can be expanded by one common form, in this case we need two numbers the sum of which is -5 and their multiplication is 6, the numbers are -2 and -3:

$$\frac{x}{(x-2)(x-3)} =$$

The rest of the process is similar to the previous example.

_ sometimes in the denominator there is powered term where we start with the least power and move toward the highest power:

$$\frac{3x-8}{x^3} = \frac{a}{x} + \frac{b}{x^2} + \frac{c}{x^3} =$$

Here x^3 is considered as the common denominator because it is dividable to all fractions:

$$\frac{a(x^2) + b(x) + c}{x^3} = \frac{ax^2 + bx + c}{x^3} =$$

At this stage we compare the obtained numerator by the original one:

There is no x^2 in the main fraction thus x^2 is multiplied by 0 abs removed thus:

a=0

X factor in the original numerator is 3 and b in the obtained one so b=3. And the constant in the original numerator is -8 thus c=-8.

The expanded fraction is as follows:

$$\frac{0}{x} + \frac{3}{x^2} - \frac{8}{x^3} = \frac{3}{x^2} - \frac{8}{x^3}$$

_The denominator of some fractions two powered terms are multiplied by each other; in order to expand these fractions we act as follows:

$$\frac{1}{x^2(x+1)^3} = \frac{a}{x} + \frac{b}{x^2} + \frac{c}{x+1} + \frac{d}{(x+1)^2} + \frac{e}{(x+1)^3}$$

Here $x^2(x+1)^3$ is considered as the common denominator and act as before.

_If the term in the parentheses in the denominator of the fraction is powered by a number we act as follows:

$$\frac{1}{(x+3)(x^2+1)} = \frac{a}{x+3} + \frac{bx+c}{x^2+1}$$

_ Note that the power of x in the numerator is one unit lower than the power of x at the denominator. Then we obtain the common denominator and act as before.

$$\frac{x^2}{x^2(x^2+5)} = \frac{a}{x} + \frac{b}{x^2} + \frac{cx+d}{x^2+5}$$

Integration by the expansion of the fractions

$$\int \frac{2x-6}{x^2-1} =$$

In the first step we expand the fraction:

The denominator can be expanded by difference/ sum of two squares:

$$\frac{2x-6}{(x-1)(x+1)} = \frac{a}{x-1} + \frac{b}{x+1} = \frac{a(x+1)+b(x-1)}{(x-1)(x+1)} = \frac{ax+a+bx-b}{(x-1)(x+1)} =$$

We factorize x in the numerator:

$$\frac{x(a+b)+a-b}{(x-1)(x+1)} =$$

Now we compare the numerator of the two fractions:

X factor in the numerator of the original fraction is 2 then a+b=2. Also the fixed number in the numerator of the original fraction is -6. Thus a-b=-6. Now we place the obtained values in the system and solve them.

$$\begin{cases} a+b=2 \\ a-b=-6 \end{cases} \rightarrow 2a = -4 \rightarrow a = \frac{-4}{2} = -2$$

$$a+b=2 \rightarrow -2+b=2 \rightarrow b = 2+2 = 4$$

The following expanded integral is obtained:

$$\int \frac{-2}{x-1} + \frac{4}{x+1} \, dx$$

$$= -2 \int \frac{1}{x-1} \, dx$$

$$+ 4 \int \frac{1}{x+1} \, dx = -2ln|x-1| + 4ln|x+1|$$

$$+ c$$

The definite integral

The definite integral is presented as $\int_a^b f(x)dx$ where $f(x)$ is the function to be integrated and a and b are the upper and lower limits.

To solve the problems of definite integral first we solve them as the indefinite integral and then obtain the upper and lower limits and deduct them.

$$\int_0^2 x^2\,dx =$$

Solution: first we solve the integral as the indefinite integral

$$\int x^2 dx = \frac{1}{3}x^3$$

Now we obtain the upper and lower limits and deduct them. To do so we replace x by 2 and 0 seperately and solve the problem.

$$\left[\frac{1}{3}(2)^3 - \frac{1}{3}(0)^3\right] = \left(\frac{1}{3} \times 8\right) - \left(\frac{1}{3} \times 0\right) = \frac{8}{3} - 0 = \frac{8}{3}$$

One might ask why the answer at the end of answer why the constant value of +c is not replaced. In response we shall say that as the name of Integral suggests the answer is a defined value do we do not need to put a hypothetical + C value.

$$\int_0^1 \frac{3x}{1+x^2}\,dx =$$

Solution: first we solve the integral as the indefinite integral

We use the method of change of variables:

Step 1: Consider $1 + x^2$ equal to u and differentiate it.

$$1 + x^2 = u \rightarrow 2x\, dx = du \rightarrow x dx = \frac{1}{2}\, du$$

Step 2: We solve the integral based on u:

$$=3 \int \frac{\frac{1}{2} du}{u} = \frac{3}{2} \int \frac{du}{u} = \frac{3}{2} ln|u|$$

Step 3: Now we replace the value of u in terms of x:

$$1 + x^2 = u \rightarrow= \frac{3}{2} ln|1 + x^2|$$

Now instead of x once we put 1 and once 0 and then deduct the obtained values:

$$\frac{3}{2} ln|1 + (1)^2| - \frac{3}{2} ln|1 + (0)^2| = \frac{3}{2} ln2 - \frac{3}{2} ln1$$
$$= \frac{3}{2} ln2 - 0 = \frac{3}{2} ln2$$

Note: $ln1$ is $log_e 1$ and we know that log 1 on any base equals 0.

End of Chapter 1

Chapter II: Further integration problems (domination)

1)$\int 6\, sin(4x - 1)\, dx =$

We use the method of change of variables:

$$4x - 1 = u \rightarrow 4dx = du \rightarrow dx = \frac{1}{4}\, du$$

$$= 6 \int sinu \times \frac{1}{4} \, du$$

$$= \frac{6}{4} \int sinu \, du = \frac{3}{2} \int sinu \, du = \frac{3}{2} \times -cosu$$

$$+ c = -\frac{3}{2} cosu + c$$

$$4x - 1 = u \rightarrow = -\frac{3}{2} cos(4x - 1) + c$$

---

2)$\int x^3 \sqrt{x^4 + 5} dx =$

We use the method of change of variables:

$$x^4 + 5 = u \rightarrow 4x^3 dx = du \rightarrow x^3 dx = \frac{1}{4} du$$

$$= \int \sqrt{u} \times \frac{1}{4} du = \frac{1}{4} \int u^{\frac{1}{2}} \, du = \frac{1}{4} \times \frac{u^{\frac{3}{2}}}{\frac{3}{2}} + c = \frac{2}{12} u^{\frac{3}{2}} + c =$$

$$\frac{1}{6} \sqrt{u^3} + c$$

$$x^4 + 5 = u \rightarrow = \frac{1}{6} \sqrt{(x^4 + 5)^3} + c$$

---

3)$\int cos^3 xsinx \, dx =$

We use the method of change of variables:

$$cos x = u \rightarrow -sin x \, dx = du \rightarrow sin x \, dx = -du$$

$$= \int u^3 \times -du = -\int u^3 du = -\frac{1}{4} u^4 + c$$

$$cosx = u \rightarrow = -\frac{1}{4} cos^4 x + c$$

---

4)$\int x \, sin(x^2 + 3) \, dx =$

We use the method of change of variables:

$$x^2 + 3 = u \rightarrow 2x\,dx = du \rightarrow x\,dx = \frac{1}{2}du$$

$$= \int \sin u \times \frac{1}{2}\,du$$

$$= \frac{1}{2}\int \sin u\,du = \frac{1}{2}$$

$$\times - \cos u + c = -\frac{1}{2}\cos u + c$$

$$x^2 + 3 = u \rightarrow = -\frac{1}{2}\cos(x^2 + 3) + c$$

5)$\int e^{\sin x} \cos x\,dx =$

We use the method of change of variables:

$$\sin x = u \rightarrow \cos x\,dx = du$$

$$= \int e^u du = e^u + c$$

$$\sin x = u \rightarrow = e^{\sin x} + c$$

6)$\int x^2 \cos(x^3)\,dx =$

We use the method of change of variables:

$$x^3 = u \rightarrow 3x^2 dx = du \rightarrow x^2 dx = \frac{1}{3}\,du$$

$$= \int \cos u \times \frac{1}{3}\,du = \frac{1}{3}\int \cos u\,du = \frac{1}{3}\sin u + c$$

$$x^3 = u \rightarrow = \frac{1}{3}\sin(x^3) + c$$

7)$\int e^t \sqrt{1 + e^t}\,dt =$

We use the method of change of variables:

$$1 + e^t = u \rightarrow e^t dt = du$$

$$= \int \sqrt{u} \, du = \int u^{\frac{1}{2}} \, du = \frac{u^{\frac{3}{2}}}{\frac{3}{2}} + c = \frac{2}{3}\sqrt{u^3} + c$$

$$1 + e^t = u \rightarrow = \frac{2}{3}\sqrt{(1+e^t)^3} + c$$

---

8)$\int \frac{1}{\sqrt{x}} \frac{1}{\left(1+2\sqrt{x}\right)^2} \, dx =$

We use the method of change of variables:

$$1 + 2\sqrt{x} = u$$

Note: 1 is a constant value and we know that the derivative of a fixed value is 0. $2\sqrt{x}$ is the derivative obtained by multiplying two terms; always remember the following formula:

$$\boxed{y = uv \rightarrow y' = u'v + v'u}$$

$$y = 2\sqrt{x} \rightarrow u = 2, u' = 0, v = \sqrt{x}, v' = \frac{1}{2\sqrt{x}}$$

$$y' = 0 \times \sqrt{x} + \frac{1}{2\sqrt{x}} \times 2 = 0 + \frac{2}{2\sqrt{x}} = \frac{1}{\sqrt{x}}$$

$$1 + 2\sqrt{x} = u \rightarrow \frac{1}{\sqrt{x}} \, dx = du$$

$$= \int \frac{1}{u^2} \times du = \int u^{-2} \, du = \frac{u^{-1}}{-1} + c = -\frac{1}{u} + c$$

$$1 + 2\sqrt{x} = u \rightarrow = -\frac{1}{1+2\sqrt{x}} + c$$

---

9)$\int \frac{\sqrt[3]{1+lnx}}{x} \, dx = \int \sqrt[3]{1 + lnx} \times \frac{1}{x} \, dx =$

We use the method of change of variables:

$$1 + lnx = u \rightarrow \frac{1}{x} dx = du$$

$$= \int \sqrt[3]{u}\, du = \int u^{\frac{1}{3}}\, du = \frac{u^{\frac{4}{3}}}{\frac{4}{3}} + c = \frac{3}{4}\sqrt[3]{u^4} + c$$

$$1 + lnx = u \rightarrow = \frac{3}{4}\sqrt[3]{(1+lnx)^4} + c$$

---

10) $\int (lnx + \frac{1}{lnx})\frac{1}{x}\, dx =$

We use the method of change of variables:

$$lnx = u \rightarrow \frac{1}{x} dx = du$$

$$= \int (u + \frac{1}{u})du = \int (\frac{u}{1} + \frac{1}{u})du = \int \frac{u^2+1}{u}\, du =$$

U is the common denominator of the fraction so it can be expanded:

$$= \int \frac{u^2}{u}\, du + \int \frac{1}{u}\, du = \int u\, du + \int \frac{du}{u} = \frac{1}{2}u^2 + ln|u| + c$$

$$lnx = u \rightarrow = \frac{1}{2}(lnx)^2 + ln|lnx| + c$$

---

11) $\int \frac{lnx}{x\sqrt{1+lnx}} dx = \int \frac{lnx}{\sqrt{1+lnx}} \times \frac{1}{x}\, dx =$

We use the method of change of variables:

$$1 + lnx = u \rightarrow \frac{1}{x} dx = du \longrightarrow lnx = u - 1$$

$$= \int \frac{u-1}{\sqrt{u}}\, du = \int \frac{u-1}{u^{\frac{1}{2}}}\, du = \int (u-1)u^{-\frac{1}{2}}\, du$$

$$= \int u^{\frac{1}{2}} - u^{-\frac{1}{2}}\, du =$$

$$\int u^{\frac{1}{2}}du - \int u^{-\frac{1}{2}}du = \frac{u^{\frac{3}{2}}}{\frac{3}{2}} - \frac{u^{\frac{1}{2}}}{\frac{1}{2}} + c = \frac{2}{3}\sqrt{u^3} - 2\sqrt{u} + c$$

$$1+lnx = u \rightarrow= \frac{2}{3}\sqrt{(1+lnx)^3} - 2\sqrt{1+lnx}+c$$

---

12)$\int \frac{e^{\frac{1}{x}}}{x^2} dx = \int e^{\frac{1}{x}} \times \frac{1}{x^2} dx =$

We use the method of change of variables:

$$\frac{1}{x} = u$$

reminder: $y = \dfrac{u}{v} \rightarrow y' = \dfrac{u'v - v'u}{v^2}$

$$y = \frac{1}{x} \rightarrow u = 1, u' = 0, v = x, v' = 1 \rightarrow y' = \frac{0 \times x - 1 \times 1}{x^2}$$

$$= -\frac{1}{x^2}$$

$$\frac{1}{x} = u \rightarrow -\frac{1}{x^2} dx = du \rightarrow \frac{1}{x^2} dx = -du$$

$$=\int e^u \times -du = -\int e^u du = -e^u+c$$

$$\frac{1}{x} = u \rightarrow= -e^{\frac{1}{x}} + c$$

---

13)$\int \frac{sin(lnx)}{x} dx = \int sin(lnx) \times \frac{1}{\square} dx =$

We use the method of change of variables:

$$lnx = u \rightarrow \frac{1}{x} dx = du$$

$$=\int sin\, u\, du = -\cos u + c \rightarrow lnx = u \rightarrow= -\cos(lnx) + c$$

---

14)$\int \frac{x\, dx}{cos^2 x^2} = \int \frac{1}{cos^2 x^2} \times x\, dx =$

We use the method of change of variables:

$$x^2 = u \rightarrow 2x\, dx = du \rightarrow x\, dx = \frac{1}{2}\, du$$

$$\boxed{\text{reminder:}\, \frac{1}{\cos^2 u} = 1 + \tan^2 u = \sec^2 u}$$

$$= \int \frac{1}{\cos^2 u} \times \frac{1}{2}\, du = \int \sec^2 u \times \frac{1}{2}\, du = \frac{1}{2}\int \sec^2 u\, du =$$

$$\frac{1}{2}\tan u + c$$

$$x^2 = u \rightarrow= \frac{1}{2}\tan(x^2) + c$$

---

15) $\int \tanh x\, dx =$

Note: always remember the following equations to integrate the hyperbolic functions:

$$\boxed{sechx = \frac{1}{\cosh x}}, \boxed{cschx = \frac{1}{\sinh x}}, \boxed{\tanh x = \frac{\sinh x}{\cosh x}},$$

$$\boxed{\coth x = \frac{\cosh x}{\sinh x}}$$

$$\boxed{1 - \tanh^2 x = \frac{1}{\cosh^2 x}}, \boxed{1 - \coth^2 x = \frac{-1}{\sinh^2 x}},$$

$$\boxed{\cosh^2 x - \sinh^2 x = 1}$$

$$\sinh^2 x = \frac{1}{2}(\cosh 2x - 1) \xrightarrow{In\ general} \sinh^2 u = \frac{1}{2}(\cosh 2u - 1)$$

$$\cosh^2 x = \frac{1}{2}(\cosh 2x + 1) \xrightarrow{In\ general} \cosh^2 u = \frac{1}{2}(\cosh 2u + 1)$$

$$= \int \frac{\sinh x}{\cosh x}\, dx =$$

We use the method of change of variables:

$Coshx = u \rightarrow \sinh x\, dx = du$

66

$$=\int \frac{du}{u} = ln|u| + c \rightarrow coshx = u \rightarrow= ln|coshx| + c$$

---

16)$\int \frac{(cosh^2x - sinh^2x)x}{1+x^2} dx = \int \frac{1 \times x}{1+x^2} dx = \int \frac{x}{1+x^2} dx =$

We use the method of change of variables:

$$1 + x^2 = u \rightarrow 2x\, dx = du \rightarrow x\, dx = \frac{1}{2} du$$

$$=\int \frac{\frac{1}{2}du}{u} = \frac{1}{2}\int \frac{du}{u} = \frac{1}{2} ln|u| + c \rightarrow 1 + x^2 = u \rightarrow= \frac{1}{2} ln|1 + x^2| + c$$

---

17)$\int sinh^2(6x + 2)dx =$

We use the method of change of variables:

$$6x + 2 = u \rightarrow 6dx = du \rightarrow dx = \frac{1}{6} du$$

$$=\int sinh^2 u \times \frac{1}{6} du = \frac{1}{6}\int sinh^2 u\, du = \frac{1}{6}\int \frac{1}{2}(cosh2u -$$
$$1)du = \frac{1}{12}\int cosh2u\, du - \frac{1}{12}\int 1\, du =$$

Note: $\int cosh2u\, du = \frac{1}{2} sinh2u + c$

$$=\frac{1}{12} \times \frac{1}{2} \times sinh2u - \frac{1}{12} u + c = \frac{1}{24} sinh2u - \frac{1}{12} u + \square$$

$$6x + 2 = u \rightarrow= \frac{1}{24} sin2(6x + 2) - \frac{1}{12}(6x + 2) + c$$

$$=\frac{1}{24} sin(12x + 4) - \frac{1}{12}(6x + 2) + c$$

---

18)$\int \frac{x^2}{\sqrt[3]{x^3+1}} dx =$

We use the method of change of variables:

$$x^3 + 1 = u \rightarrow 3x^2 dx = du \rightarrow x^2 dx = \frac{1}{3} du$$

$$=\int \frac{\frac{1}{3}du}{\sqrt[3]{u}} = \frac{1}{3}\int \frac{du}{u^{\frac{1}{3}}} = \frac{1}{3}\int u^{-\frac{1}{3}} du = \frac{1}{3} \times \frac{u^{\frac{2}{3}}}{\frac{2}{3}} + c = \frac{1}{2}\sqrt[3]{u^2} + c$$

$$x^3 + 1 = u \rightarrow= \frac{1}{2}\sqrt[3]{(x^3 + 1)^2} + c$$

---

19) $\int \sqrt{\frac{arcsinx}{1-x^2}}\, dx =$

Reminder→ $\boxed{\sqrt[n]{\frac{a}{b}} = \frac{\sqrt[n]{a}}{\sqrt[n]{b}}}$

$$=\int \frac{\sqrt{arcsinx}}{\sqrt{1-x^2}}\, dx = \int \sqrt{arcsinx} \times \frac{1}{\sqrt{1-x^2}}\, dx =$$

We use the method of change of variables:

$$arcsinx = u \rightarrow \frac{1}{\sqrt{1-x^2}}\, dx = du$$

$$=\int \sqrt{u}\, du = \int u^{\frac{1}{2}}\, du = \frac{u^{\frac{3}{2}}}{\frac{3}{2}} + c = \frac{2}{3}\sqrt{u^3} + c$$

$$arcsinx = u \rightarrow= \frac{2}{3}\sqrt{(arcsinx)^3} + c$$

---

20) $\int \frac{arctanx}{1+x^2}\, dx = \int arctanx \times \frac{1}{1+x^2}\, dx =$

We use the method of change of variables:

$$arctanx = u \rightarrow \frac{1}{1+x^2}\, dx = du$$

$$=\int u\, du = \frac{1}{2}u^2 + c \rightarrow arctanx = u \rightarrow= \frac{1}{2}(arctanx)^2 + c$$

---

21) $\int \frac{1+lnx}{3+xlnx}\, dx =$

We use the method of change of variables:

$$3 + xlnx = u$$

Note: 3 is constant and constant value derivative is zero, also xlnx is the derivative of the multiplication of two terms then:

$$y = uv \rightarrow y' = u'v + v'u \rightarrow y = xlnx \rightarrow x = u \rightarrow 1 = u', lnx$$

$$= v \rightarrow \frac{1}{x} = v'$$

$$y' = 1 \times lnx + \frac{1}{x} \times x = lnx + \frac{x}{x} = lnx + 1$$

Then→$1 + lnx\ dx = du$

$$=\int \frac{du}{u} = ln|u| + c = ln|3 + lnx| + c$$

---

22)$\int \frac{sinx}{\sqrt[5]{cosx}} dx =$

We use the method of change of variables:

$$cosx = u \rightarrow -sinx\ dx = du \rightarrow sinx\ dx = -du$$

$$=\int \frac{-du}{\sqrt[5]{u}} = -\int \frac{du}{u^{\frac{1}{5}}} = -\int u^{-\frac{1}{5}} du = -\frac{u^{\frac{4}{5}}}{\frac{4}{5}} + c = -\frac{5}{4}\sqrt[5]{u^4} + c$$

$$cosx = u \rightarrow= -\frac{5}{4}\sqrt[5]{cos^4x} + c$$

Further explanation: As mentioned before, in dividing the fractions we use cross multiplication, here in order to obtain the result of $\frac{u^{\frac{4}{5}}}{\frac{4}{5}}$ as fraction i.e. $u^{\frac{4}{5}}$ we use cross multiplication and multiply $u^{\frac{4}{5}}$ abd 1 by 5 and 4 respectively.

---

23)$\int ln(x + 1)\ dx =$

We use the method of change of variables:

$x + 1 = u \rightarrow dx = du$

$= \int \ln u \ du = u \ln u - u + c \rightarrow x + 1 = u \rightarrow = (x + 1) \ln(x + 1) - (x + 1) + c$

Note: $\int \ln x \ dx = x \ln x - x + c \xrightarrow{In\ general} \int \ln u \ du = u \ln u - u + c$

---

24)$\int e^x \sqrt{9 + e^x} \ dx =$

We use the method of change of variables:

$9 + e^x = u \rightarrow e^x \ dx = du$

$= \int \sqrt{u} \ du = \int u^{\frac{1}{2}} \ du = \frac{u^{\frac{3}{2}}}{\frac{3}{2}} + c \rightarrow = \frac{2}{3}\sqrt{u^3} + c \rightarrow 9 + e^x = u \rightarrow$

$= \frac{2}{3}\sqrt{(9 + e^x)^3} + c$

---

25)$\int \frac{\cos 5x}{e^{\sin 5x}} \ dx =$

We use the method of change of variables:

$\sin 5x = u \rightarrow 5\cos 5x \ dx = du \rightarrow \cos 5x \ dx = \frac{1}{5} \ du$

$= \int \frac{\frac{1}{5} du}{e^u} = \frac{1}{5} \int \frac{du}{e^u} = \frac{1}{5} \int e^{-u} \ du = \frac{1}{5} \times -e^{-u} + c = -\frac{1}{5}e^{-u} + c$

$\sin 5x = u \rightarrow = -\frac{1}{5}e^{-\sin 5x} + c$

$\boxed{\text{Note: } \int e^u du = e^u + c, \int e^{-u} du = -e^{-u} + c}$

---

26)$\int e^x(1 + 2e^x)^4 dx =$

We use the method of change of variables:

$1 + 2e^x \rightarrow 2e^x dx = du \rightarrow e^x dx = \frac{1}{2} du$

$$= \int u^4 \times \frac{1}{2} \, du = \frac{1}{2} \int u^4 du = \frac{1}{2} \times \frac{u^5}{5} + c = \frac{u^5}{10} + c$$

$$1 + 2e^x \rightarrow = \frac{(1 + 2e^x)^5}{10} + c$$

---

27) $\int \frac{2x+1}{3x^2+3x-2} \, dx =$

We use the method of change of variables:

$$3x^2 + 3x - 2 = u \rightarrow 6x + 3 \, dx = du$$

If we divide both sides by 3 we have:

$$2x + 1 \, dx = \frac{1}{3} du$$

$$= \int \frac{\frac{1}{3}du}{u} = \frac{1}{3} \int \frac{du}{u} = \frac{1}{3} ln|u| + c \rightarrow 3x^2 + 3x - 2 = u \rightarrow =$$
$$\frac{1}{3} ln|3x^2 + 3x - 2| + c$$

28) $\int \frac{arcsecx}{x\sqrt{x^2-1}} \, dx = \int arcsecx \times \frac{1}{x\sqrt{x^2-1}} dx =$

We use the method of change of variables:

$$arcsecx = u \rightarrow \frac{1}{x\sqrt{x^2 - 1}} \, dx = du$$

$$= \int u \, du = \frac{1}{2}u^2 + c \rightarrow arcsecx = u \rightarrow = \frac{1}{2}(arcsecx)^2 + c$$

---

29) $\int \frac{1}{\sqrt{x}} \frac{1}{1+\sqrt{x}} dx =$

We use the method of change of variables:

$$\sqrt{x} = u \rightarrow \frac{1}{2\sqrt{x}} \, dx = du \rightarrow \frac{1}{\sqrt{x}} \, dx = 2du$$

$$= \int \frac{1}{1+u} \times 2du = 2 \int \frac{du}{1+u} =$$

We use the method of change of variables gain:

$1 + u = z \rightarrow du = dz$

$=2\int \frac{dz}{z} = 2ln|z| + c \rightarrow 1 + u = z \rightarrow= 2ln|1 + u| + c \rightarrow$

$\sqrt{x} = u \rightarrow= 2ln|1 + \sqrt{x}| + c$

---

30)$\int xsin(x^2 + 1)dx =$

We use the method of change of variables:

$x^2 + 1 = u \rightarrow 2x\, dx = du \rightarrow x\, dx = \frac{1}{2}\, du$

$=\int sin\, u \times \frac{1}{2}\, du = \frac{1}{2}\int sinu\, du = \frac{1}{2} \times -cosu + c = -\frac{1}{2}cosu + c$

$x^2 + 1 = u \rightarrow= -\frac{1}{2}cos(x^2 + 1) + c$

---

31)$\int x^{-3}\sqrt{3 + 5x^{-2}}\, dx =$

We use the method of change of variables:

$3 + 5x^{-2} = u \rightarrow -10x^{-3}dx = du \rightarrow x^{-3}dx = -\frac{1}{10}\, du$

$=\int \sqrt{u} \times -\frac{1}{10}\, du = -\frac{1}{10}\int u^{\frac{1}{2}}\, du = -\frac{1}{10} \times \frac{u^{\frac{3}{2}}}{\frac{3}{2}} + c = -\frac{1}{15}u^{\frac{3}{2}} + c$

$c = -\frac{1}{15}\sqrt{u^3} + c$

$3 + 5x^{-2} = u \rightarrow= -\frac{1}{15}\sqrt{(3 + 5x^{-2})^3} + c$

---

32)$\int \sqrt{x}\cos(x\sqrt{x})\, dx =$

$\sqrt{x} = u \rightarrow x = u^2 \rightarrow dx = 2u\, du$

$=\int \cos(u^2 \times u) \times u \times 2u\, du = \int \cos(u^3) \times 2u^2 du =$
$2\int \cos(u^3)u^2\, du =$

We use the method of change of variables:

$$u^3 = z \rightarrow 3u^2 du = dz \rightarrow u^2 du = \frac{1}{3}dz$$

$$=2\int \cos z \times \frac{1}{3}dz = \frac{2}{3}\int \cos z \, dz = \frac{2}{3}sinz + c$$

$$u^3 = z \rightarrow= \frac{2}{3}sin(u^3) + c \rightarrow \sqrt{x} = u \rightarrow= \frac{2}{3}sin(\sqrt{x})^3 + c$$

33)$\int (e^x + 1)^2 dx =$

We use the method of change of variables:

$$e^x = u \rightarrow e^x dx = du \rightarrow dx = \frac{1}{e^x}du \rightarrow dx = \frac{1}{u}du$$

$$=\int (u + 1)^2 \times \frac{1}{u}du =$$

$(u + 1)^2$ is full square thus:

$$=\int (u^2 + 2u + 1) \times \frac{1}{u}du =$$

Now we multiply all terms in the parentheses by $\frac{1}{u}$:

$$=\int (\frac{u^2}{u} + \frac{2u}{u} + \frac{1}{u})du = \int (u + 2 + \frac{1}{u})du = \int u du +$$
$$2 \int du + \int \frac{1}{u}du =$$

$$=\frac{1}{2}u^2 + 2u + ln|u| + c \rightarrow u = e^x \rightarrow= \frac{1}{2}(e^x)^2 + 2(e^x) +$$
$$ln|e^x| + x$$

$$=\frac{1}{2}e^{2x} + 2e^x + ln|e^x| + c = \frac{1}{2}e^{2x} + 2e^x + x + c$$

Note: $log_a x^n = n \, log_a x$

Note 2:$log_{a^m} x^n = \frac{n}{m} log_a x$

Example 1: $log_2 4 = log_2 2^2 = 2 \, log_2 2 = 2 \times 1 = 2$

Example 2: $log_4 8 = log_{2^2} 2^3 = \frac{3}{2} log_2 2 = \frac{3}{2} \times 1 = \frac{3}{2}$

Example 3: $lne^x = log_e e^x = x \, log_e e = x \times 1 = x$

---

34) $\int \frac{x \, dx}{(2x^2+1)^3} =$

We use the method of change of variables:

$$2x^2 + 1 = u \rightarrow 4x \, dx = du \rightarrow x \, dx = \frac{1}{4} du$$

$$= \int \frac{\frac{1}{4} du}{u^3} = \frac{1}{4} \int u^{-3} \, du = \frac{1}{4} \times \frac{u^{-2}}{-2} + c = -\frac{1}{8} u^{-2} + c = -\frac{1}{8u^2} + c$$

$$2x^2 + 1 = u \rightarrow = -\frac{1}{8(2x^2 + 1)^2} + c$$

---

35) $\int (6x + 1)\sqrt{2x^3 + x + 7} \, dx =$

We use the method of change of variables:

$$2x^3 + x + 7 = u \rightarrow 6x + 1 \, dx = du$$

$$= \int \sqrt{u} \, du = \int u^{\frac{1}{2}} \, du = \frac{u^{\frac{3}{2}}}{\frac{3}{2}} + c = \frac{2}{3} \sqrt{u^3} + c$$

$$2x^3 + x + 7 = u \rightarrow = \frac{2}{3} \sqrt{(2x^3 + x + 7)^3} + c$$

---

36) $\int \frac{1}{xlnxlnlnx} \, dx = \int \frac{1}{lnxlnlnx} \times \frac{1}{x} dx$

We use the method of change of variables:

$$lnx = u \rightarrow \frac{1}{x} dx = du$$

$$= \int \frac{1}{ulnu} \, du =$$

We use the method of change of variables again:

74

$$lnu = z \to \frac{1}{u} du = dz$$

$$= \int \frac{1}{z} dz = ln|z| + c \to z = lnu \to= ln|lnu| + c \to u = lnx \to$$
$$= ln|ln(lnx)| + c$$

---

37)$\int \frac{cos(2x)}{1+sin(2x)} dx =$

We use the method of change of variables:

Reminder: $\qquad y = sinu \to y' = u'cosu, y = cosu \to y' = -u'sinu$

$$1 + sin(2x) = u \to 2cos2x \, dx = du \to cos2x \, dx = \frac{1}{2} du$$

$$= \int \frac{\frac{1}{2}du}{u} = \frac{1}{2} \int \frac{du}{u} = \frac{1}{2} ln|u| + c \to 1 + sin(2x) = u \to=$$
$$\frac{1}{2} ln|1 + sin(2x)| + c$$

---

38)$\int sec^2 \, xtanx \, dx =$

We use the method of change of variables:

$$tanx = u \to 1 + tan^2x \, dx = du$$

Reminder: $1 + tan^2x = \frac{1}{cos^2x} = sec^2x$

$$sec^2x \, dx = du$$

$$= \int u \, du = \frac{1}{2} u^2 + c \to tanx = u \to= \frac{1}{2} (tanx)^2 + c$$

---

39)$\int e^{tanx} \, sec^2x \, dx =$

We use the method of change of variables:

$$tanx = u \to 1 + tan^2x \, dx = du \to sec^2x \, dx = du$$

$$= \int e^u \, du = e^u + c \to tanx = u \to= e^{tanx} + c$$

---

40) $\int \frac{x^3}{1+x^8} \, dx =$

Reminder: $(a^n)^m = a^{n \times m} \longrightarrow x^8 = (x^4)^2$

$= \int \frac{x^3}{1+(x^4)^2} \, dx =$

We use the method of change of variables:

$$x^4 = u \rightarrow 4x^3 dx = du \rightarrow x^3 dx = \frac{1}{4} \, du$$

$$= \int \frac{\frac{1}{4} du}{1+u^2} = \frac{1}{4} \int \frac{du}{1+u^2} = \frac{1}{4} \arctan u + c$$

$$x^4 = u \rightarrow = \frac{1}{4} \arctan x^4 + c$$

---

41) $\int \frac{\cot x}{\sin^2 x} \, dx = \int \cot x \times \frac{1}{\sin^2 x} \, dx =$

We use the method of change of variables:

$$\cot x = u \rightarrow -(1 + \cot^2 x) dx = du$$

Reminder: $1 + \cot^2 x = \frac{1}{\sin^2 x} = \csc^2 x$

$$\rightarrow -\frac{1}{\sin^2 x} \, dx = du \rightarrow \frac{1}{\sin^2 x} \, dx = -du$$

$$= \int u \times -du = -\int u \, du = -\frac{1}{2} u^2 + c \rightarrow \cot x = u \rightarrow =$$

$$-\frac{1}{2} \cot^2 x + c$$

---

42) $\int \frac{\tan \sqrt{x+2}}{\sqrt{x+2}} \, dx =$

We use the method of change of variables:

$$\sqrt{x+2} = u$$

Reminder: $\sqrt[m]{u^n} = u^{\frac{n}{m}} \rightarrow \sqrt{x+2} = (x+2)^{\frac{1}{2}}$

$$y = (x+2)^{\frac{1}{2}} \rightarrow y' = \frac{1}{2}(x+2)^{-\frac{1}{2}} = \frac{1}{2} \times \frac{1}{(x+2)^{\frac{1}{2}}} = \frac{1}{2\sqrt{x+2}}$$

$$\frac{1}{2\sqrt{x+2}}dx = du \rightarrow \frac{1}{\sqrt{x+2}}dx = 2du$$

$$=\int tanu \times 2du = 2\int tanu \, du = 2 \times -ln|cosu| + c =$$
$$-2ln|cosu| + c$$

$$\sqrt{x+2} = u \rightarrow= -2ln\left|cos\sqrt{x+2}\right| + c$$

---

43)$x^2(3-10x^3)^4dx =$

We use the method of change of variables:

$$3 - 10x^3 = u \rightarrow -30x^2dx = du \rightarrow x^2dx = -\frac{1}{30}du$$

$$=\int u^4 \times -\frac{1}{30}du = -\frac{1}{30}\int u^4 \, du = -\frac{1}{30} \times \frac{u^5}{5} + c = -\frac{1}{150}u^5 +$$
$$c$$

$$3 - 10x^3 = u \rightarrow= -\frac{1}{150}(3-10x^3)^5 + c$$

---

44)$\int \frac{e^x}{e^x+e^{-x}}dx =$

We use the method of change of variables:

$$e^x = u \rightarrow e^xdx = du \rightarrow e^{-x} = \frac{1}{e^x} = \frac{1}{u}$$

$$=\int \frac{du}{u+\frac{1}{u}} = \int \frac{du}{\frac{u^2+1}{u}} = \int \frac{\frac{du}{1}}{\frac{u^2+1}{u}} = \int \frac{udu}{1+u^2} =$$

We use the method of change of variables again:

$$1 + u^2 = z \rightarrow 2u \, du = dz \rightarrow u \, du = \frac{1}{2}dz$$

$$=\int \frac{\frac{1}{2}dz}{z} = \frac{1}{2}\int \frac{dz}{z} = \frac{1}{2}ln|z| + c \rightarrow 1 + u^2 = z \rightarrow = \frac{1}{2}ln|1 + u^2| + c$$

$$e^x = u \rightarrow = \frac{1}{2}ln|1 + (e^x)^2| + c = \frac{1}{2}ln|1 + e^{2x}| + c$$

---

45)$\int \frac{e^{2x}}{e^x-7}dx = \int \frac{e^x \times e^x}{e^x-7}dx =$

We use the method of change of variables:

$$e^x - 7 = u \rightarrow e^x dx = du \rightarrow e^x = u + 7$$

$$=\int \frac{(u+7)du}{u} = \int \frac{u}{u}du + \int \frac{7}{u}du = \int 1 \, du + 7\int \frac{du}{u} = u +$$
$$7ln|u| + c$$

$$e^x - 7 = u \rightarrow = (e^x - 7) + 7ln|e^x - 7| + c$$

---

46)$\int \sqrt{e^x} \, dx = \int e^{\frac{x}{2}} \, dx =$

We use the method of change of variables:

$$\frac{1}{2}x = u \rightarrow \frac{1}{2}dx = du \rightarrow dx = 2du$$

$$=\int e^u \times 2du = 2\int e^u \, du = 2e^u + c \rightarrow \frac{1}{2}x = u \rightarrow = 2e^{\frac{x}{2}} +$$
$$c = 2\sqrt{e^x} + c$$

---

47)$\int \frac{e^{-3x}}{1-e^{-3x}}dx =$

We use the method of change of variables:

Reminder:$y = e^u \rightarrow y' = u'e^u$

$$1 - e^{-3x} = u \rightarrow -(-3)e^{-3x}dx = du \rightarrow 3e^{-3x}dx = du$$
$$\rightarrow e^{-3x}dx = \frac{1}{3}du$$

$$=\int \frac{\frac{1}{3}du}{u} = \frac{1}{3}\int \frac{du}{u} = \frac{1}{3} \times ln|u| + c \to 1 - e^{-3x} = u \to=$$

$$\frac{1}{3}ln|1 - e^{-3x}| + c$$

---

48)$\int \frac{cot(lnx)}{x} dx = \int cot(lnx) \times \frac{1}{x} dx$

We use the method of change of variables:

$$lnx = u \to \frac{1}{x}dx = du$$

$$=\int cot\, u\; du = \int \frac{cosu}{sinu} du =$$

We use the method of change of variables again:

$$sinu = z \to cos\, u\; du = dz$$

$$=\int \frac{dz}{z} = ln|z| + c \to sinu = z \to= ln|sinu| + c \to$$

$$lnx = u \to= ln|sin(lnx)| + c$$

---

49)$\int sin^2(3x)dx =$

We use the method of change of variables:

$$3x = u \to 3dx = du \to dx = \frac{1}{3} du$$

$$=\int sin^2 u \times \frac{1}{3}du = \frac{1}{3}\int sin^2 u\; du =$$

Reminder: $sin^2 u = \frac{1-cos2u}{2}$

$$=\frac{1}{3}\int \frac{1-cos2u}{2} du = \frac{1}{3}\int \frac{1}{2} - \frac{1}{2}cos2u\; du = \frac{1}{6}\int du -$$

$$\frac{1}{6}\int cos2u\; du =$$

Note: $\int cosau\; du = \frac{1}{a}sinau + c, \int sinau\; du = -\frac{1}{a}cosau + c$

In the above formula a is constant:

$$=\frac{1}{6}u - \frac{1}{6}\times\frac{1}{2}sin2u + c = \frac{1}{6}u - \frac{1}{12}sin2u + c \rightarrow 3x = u \rightarrow=$$

$$\frac{1}{6}\times 3x - \frac{1}{12}sin2(3x) + c = \frac{1}{2}x - \frac{1}{12}sin6x + c$$

---

50)$\int \frac{4x+8}{x^2+4x-25} dx =$

We use the method of change of variables:

$x^2 + 4x - 25 = u \rightarrow 2x + 4\ dx = du$

We need $4x + 8$ so we multiply both sides by 2:

$4x + 8\ dx = 2du$

$$=\int \frac{2\ du}{u} = 2\int \frac{du}{u} = 2ln|u| + c \rightarrow x^2 + 4x - 25 = u \rightarrow=$$
$$2ln|x^2 + 4x - 25| + c$$

---

51)$\int \frac{coshx}{2+3sinhx} dx =$

We use the method of change of variables:

$$2 + 3sinhx = u \rightarrow 3coshx\ dx = du \rightarrow coshx\ dx = \frac{1}{3}du$$

$$=\int \frac{\frac{1}{3}du}{u} = \frac{1}{3}\int \frac{du}{u} = \frac{1}{3}ln|u| + c \rightarrow 2 + 3sinhx = u \rightarrow=$$
$$\frac{1}{3}ln|2 + 3sinhx| + c$$

---

52)$\int \frac{dx}{x^2+4x+8} =$

8 can be written as 4+4 then:

$$=\int \frac{dx}{x^2+4x+4+4} =$$

Note:$(a + b)^2 = a^2 + 2ab + b^2$

Example) $(x + 2)^2 = x^2 + 4x + 4$

$$=\int \frac{dx}{(x+2)^2+4} =$$

We use the method of change of variables:

$$x + 2 = u \rightarrow dx = du$$

$$= \int \frac{du}{u^2+4} =$$

Note: $\int \frac{du}{a^2+u^2} = \frac{1}{a} \arctan \frac{u}{a} + c$

$$= \int \frac{du}{u^2+2^2} = \frac{1}{2} \arctan \frac{u}{2} + c \rightarrow x + 2 = u \rightarrow = \frac{1}{2} \arctan \frac{x+2}{2} + c$$

---

53) $\int \frac{\cos\sqrt{x}}{\sqrt{x}} dx =$

We use the method of change of variables:

$$\sqrt{x} = u \rightarrow \frac{1}{2\sqrt{x}} dx = du \rightarrow \frac{1}{\sqrt{x}} dx = 2du$$

$$= \int \cos u \times 2 du = 2 \int \cos u \, du = 2 \sin u + c$$

$$\sqrt{x} = u \rightarrow = 2 \sin(\sqrt{x}) + c$$

---

54) $\int \sec^4 x \sec x \tan x \, dx =$

We use the method of change of variables:

$$\sec x = u \rightarrow \sec x \tan x \, dx = du$$

$$= \int u^4 \, du = \frac{1}{5} u^5 + c \rightarrow \sec x = u \rightarrow = \frac{1}{5} \sec^5 x + c$$

---

55) $\int 3\cos^2(5x) dx = 3 \int \cos^2(5x) dx =$

We use the method of change of variables:

$$5x = u \rightarrow 5dx = du \rightarrow dx = \frac{1}{5} du$$

$$= 3 \int \cos^2 u \times \frac{1}{5} du = \frac{3}{5} \int \cos^2 u \, du =$$

Reminder: $\cos^2 u = \frac{1+\cos 2u}{2}$

$$\frac{-3}{5}\int \frac{1+cos2u}{2} = \frac{3}{5}\int \frac{1}{2}du + \frac{3}{5}\int \frac{1}{2}cos2u \ du = \frac{3}{10}\int du +$$

$$\frac{3}{10}\int cos2u \ du=$$

Note: $\int cosau \ du = \frac{1}{a}sinau + c, \int sinau \ du = -\frac{1}{a}cosau + c$

$$=\frac{3}{10}u + \frac{3}{10} \times \frac{1}{2}sin2u + c$$

$$5x = u \rightarrow= \frac{3}{10}(5x) + \frac{3}{20}sin2(5x) + c$$

$$= \frac{15x}{10} + \frac{3sin10x}{20} + c = \frac{3}{2}x + \frac{3}{20}sin10x + c$$

---

56)$\int (sec^2 x)\sqrt{5 + tanx} \ dx =$

We use the method of change of variables:

$5 + tanx = u \rightarrow 1 + tan^2 x \ dx = du \rightarrow sec^2 x \ dx = du$

$$=\int \sqrt{u} \ du = \int u^{\frac{1}{2}} dy = \frac{u^{\frac{3}{2}}}{\frac{3}{2}} + c = \frac{2}{3}u^{\frac{3}{2}} + c = \frac{2}{3}\sqrt{u^3} + c$$

$$5 + tanx = u \rightarrow= \frac{2}{3}\sqrt{(5 + tanx)^3} + c$$

---

57)$\int \frac{cosxln(sinx)}{sinx}dx = \int \ln(sinx) \times \frac{cosx}{sinx} \ dx =$

We use the method of change of variables:

$ln(sinx) = u$

Note:$y = lnu \rightarrow y' = \frac{u'}{u}$

$ln(sinx) = lnu \rightarrow sinx = u \rightarrow cosx = u'$

$$\frac{cosx}{sinx}dx = du$$

$$=\int u \ du = \frac{1}{2}u^2 + c \rightarrow \ln(sinx) = u \rightarrow= \frac{1}{2}(ln(sinx))^2 + c$$

---

58) $\int \cos x\, e^{4+\sin x}\, dx =$

We use the method of change of variables:

$4 + \sin x = u \rightarrow \cos x\, dx = du$

$= \int e^u\, du = e^u + c \rightarrow 4 + \sin x = u \Rightarrow e^{4+\sin x} + c$

---

59) $\int (\sec x \tan x)\sqrt{4 + 3\sec x}\; dx =$

We use the method of change of variables:

$4 + 3\sec x = u \rightarrow 3\sec x \tan x\, dx = du \rightarrow \sec x \tan x\, dx$

$$= \frac{1}{3}\,du$$

$= \int \sqrt{u} \times \frac{1}{3}\,du = \frac{1}{3}\int u^{\frac{1}{2}}du = \frac{1}{3} \times \frac{u^{\frac{3}{2}}}{\frac{3}{2}} + c = \frac{2}{9}u^{\frac{3}{2}} + c = \frac{2}{9}\sqrt{u^3} +$

c

$4 + 3\sec x = u \Rightarrow \dfrac{2}{9}\sqrt{(4 + 3\sec x)^3} + c$

---

60) $\int \frac{\sin 2x - \cos 2x}{\sin 2x + \cos 2x}\, dx =$

We use the method of change of variables:

$\sin 2x + \cos 2x = u \rightarrow 2\cos 2x - 2\sin 2x\, dx = du$

$2(\cos 2x - \sin 2x)dx = du \rightarrow -2(\sin 2x - \cos 2x)dx = du$

$$\rightarrow \sin 2x - \cos 2x\, dx = -\frac{1}{2}\,du$$

$= \int \frac{-\frac{1}{2}}{u}\,du = -\frac{1}{2}\int \frac{du}{u} = -\frac{1}{2}\ln|u| + c \rightarrow \sin 2x + \cos 2x = u \rightarrow$

$= -\frac{1}{2}\ln|\sin 2x + \cos 2x| + c$

---

61) $\int \frac{\sin x + \cos x}{e^{-x} + \sin x}\, dx =$

As you can see, all the terms in question, except e^{-x} are trigonometric thus e^{-x} is an annoying term that must be removed. We know that a term can be multiplied by a term and divided by it without causing any change. Thus:

$$= \int \frac{sinx+cosx}{e^{-x}+sinx} \times \frac{e^x}{e^x} dx = \int \frac{e^x sinx+e^x cosx}{e^{-x}\times e^x+e^x sinx} dx =$$

Note:$a^n \times a^m = a^{n+m} \longrightarrow e^{-x} \times e^x = e^{-x+x} = e^0 = 1$

We know that the result of any number powered by 0 is 1.

$$= \int \frac{e^x sinx+e^x cosx}{1+e^x sinx} dx =$$

We use the method of change of variables:

$$1 + e^x sinx = u \rightarrow e^x sinx + e^x cosx\ dx = du$$

$$= \int \frac{du}{u} = ln|u| + c \rightarrow 1 + e^x sinx = u \rightarrow= ln|1 + e^x sinx| + c$$

---

62)$\int \frac{\sqrt{9-x^2}}{x^2} dx = -\frac{\pi}{2} \ll \theta \ll \frac{\pi}{2}$

We use the method of trigonometric variables:

First we factorize 9 under the radical:

$$= \int \frac{\sqrt{9(1-\frac{x^2}{9})}}{x^2} dx =$$

The term $1 - \frac{x^2}{9}$ reminds us about $1 - sin^2\theta$ thus:

$$1 - \frac{x^2}{9} = 1 - sin^2\theta \rightarrow \frac{x^2}{9} = sin^2\theta \rightarrow \frac{x}{3} = sin\theta$$

We divide $sin\theta$ by 1 and use cross multiplication:

$$\frac{x}{3} = \frac{\sin \theta}{1} \rightarrow x = 3 \sin \theta \rightarrow dx = 3\cos\theta . d\theta$$

To obtain θ we divide the known factor with unknown one:

$$\theta = \frac{x}{3\sin} = \frac{x}{3} \times \frac{1}{\sin} = arcsin(\frac{x}{3})$$

At this stage, the obtained values are placed in the first integral:

$$= \int \frac{\sqrt{9(1-sin^2\theta)}}{9sin^2\theta} \times 3\cos\theta . d\theta = \int \frac{\sqrt{9} \times\sqrt{1-sin^2\theta}}{9sin^2\theta} \times 3\cos\theta . d\theta$$

$$= \int \frac{3\times\sqrt{cos^2\theta}}{9sin^2\theta} \times 3\cos\theta . d\theta=$$

Note:$\sqrt{cos^2\theta} = |\cos\theta|$

Note: Here since the range of choice is $-\frac{\pi}{2} \ll \theta \ll \frac{\pi}{2}$ and we know that cosine is always positive within this range there is no need for absolute sign.

$$= \int \frac{9cos^2\theta}{9sin^2\theta} d\theta = \int \frac{cos^2\theta}{sin^2\theta} d\theta = \int cot^2 \theta d\theta = \int csc^2 \theta - 1 d\theta = \int csc^2 \theta d\theta - \int 1 d\theta$$

$$= -cot\theta - \theta + c$$

Consider the following triangle:

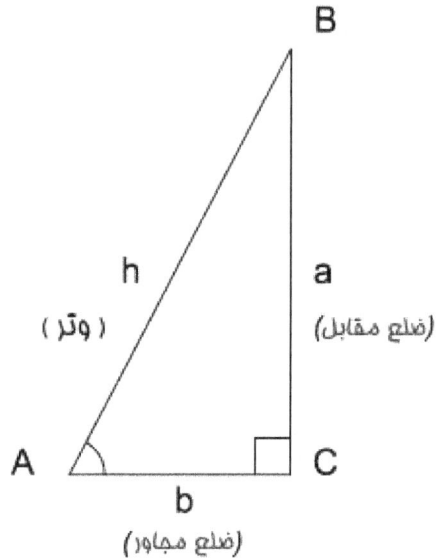

B

h
(وتر)

a
(ضلع مقابل)

A

b
(ضلع مجاور)

C

$$\frac{x}{3} = sin\ \theta = \frac{opposite\ side}{hypotenuse} = \frac{a}{h} \rightarrow a = x, 3 = h$$

Also, according to the Pythagorean Theorem hypotenuse (h) powered by 2 equals the some of other sides powered by 2.

$$3^2 = a^2 + b^2 \rightarrow x = a \rightarrow x^2 + b^2 = 9$$

$$b^2 = 9 - x^2 \rightarrow b = \sqrt{9 - x^2}$$

$$cot\theta = \frac{adjacent\ side}{opposite\ side} = \frac{b}{a} = \frac{\sqrt{9 - x^2}}{x}$$

The final answer: $=-\frac{\sqrt{9-x^2}}{x} - arcsin(\frac{x}{3}) + c$

---

62)62)$\int \frac{1}{3+x^2} dx =$

We use the method of trigonometric variables:

Method 1) the method of changing trigonometric variables:

If we factorize 3 at the denominator, the integral will be as follows:

$$\int \frac{1}{3(1+\frac{x^2}{3})} dx =$$

The term $1+\frac{x^2}{3}$ reminds us about $1+tan^2\theta$ thus:

$$1 + \frac{x^2}{3} = 1 + tan^2\theta \rightarrow \frac{x^2}{3} = tan^2\theta \xrightarrow{\sqrt{\ }} \frac{x}{\sqrt{3}} = tan\theta$$

We divide $tan\theta$ by 1 and use cross multiplication:

$$\frac{x}{\sqrt{3}} = \frac{tan\theta}{1} \rightarrow x = \sqrt{3}tan\theta \rightarrow dx = \sqrt{3}sec^2\theta.d\theta$$

$$\theta = \frac{\frac{x}{\sqrt{3}}}{\frac{tan}{1}} = \frac{x \times 1}{\sqrt{3} \times tan} = arctan\frac{x}{\sqrt{3}}$$

$$=\int \frac{\sqrt{3} \times sec^2\theta.d\theta}{3(1+tan^2\theta)} = \frac{\sqrt{3}sec^2\theta.d\theta}{3sec^2\theta} = \frac{\sqrt{3}}{3}\int \frac{sec^2\theta.d\theta}{sec^2\theta} =$$

$$\frac{\sqrt{3}}{3} = \frac{1}{\sqrt{3}} \xrightarrow{prove} \frac{1}{\sqrt{3}} \times \frac{\sqrt{3}}{\sqrt{3}} = \frac{\sqrt{3}}{\sqrt{9}} = \frac{\sqrt{3}}{3}$$

$$=\frac{1}{\sqrt{3}}\int 1 \, d\theta = \frac{1}{\sqrt{3}}\theta + c$$

Now we replace the value of θ in terms of x:

$$=\frac{1}{\sqrt{3}} arc \, tan\frac{x}{\sqrt{3}} + c$$

Second method: a quick method using the formula:

$$\cdot\int \frac{1}{a^2+x^2} dx = \frac{1}{a} arctan\frac{x}{a} + c$$

$$a = \sqrt{3} \to = \frac{1}{\sqrt{3}} arctan \frac{x}{\sqrt{3}} + c$$

---

63) Prove: $\int \frac{1}{\sqrt{a^2 - x^2}} dx = arcsin \frac{x}{a} + c - \frac{\pi}{2} \ll \theta \ll \frac{\pi}{2}$

We use the method of trigonometric variables:

We factorize a^2 in the denominator and under the radical:

$$\int \frac{1}{\sqrt{a^2 (1 - \frac{x^2}{a^2})}} dx =$$

The term $1 - \frac{x^2}{a^2}$ reminds us about $1 - sin^2 \theta$ thus:

$$1 - \frac{x^2}{a^2} = 1 - sin^2 \theta \to \frac{x^2}{a^2} = sin^2 \theta \to \frac{x}{a} = sin\,\theta \to x$$
$$= a\,sin\,\theta \to dx = acos\theta . d\theta$$

$$\theta = \frac{\frac{x}{a}}{\frac{sin}{1}} = \frac{x \times 1}{asin} = \frac{x}{asin} = \frac{x}{a} \times \frac{1}{sin} = arcsin \frac{x}{a}$$

$$= \int \frac{acos\theta . d\theta}{\sqrt{a^2 (1 - sin^2 \theta)}} = \int \frac{acos\theta . d\theta}{\sqrt{a^2} \times \sqrt{cos^2 \theta}} = \int \frac{acos\theta . d\theta}{acos\theta} = \int 1 d\theta = \theta + c$$

Now we replace the value of θ in terms of x:

$$= arcsin \frac{x}{a} + c$$

---

64) $\int \frac{lnx}{x^5} dx = \int lnx \times \frac{1}{x^5} dx =$

We use fractional method:

$$lnx = u \to \frac{1}{x} dx = du, \frac{1}{x^5} dx = dv \to x^{-5} dx = dv \to \frac{x^{-4}}{-4} = v$$

$$\to -\frac{1}{4x^4} = v$$

$$\int u\,dv = uv - \int v\,du = lnx \times -\frac{1}{4x^4} - \int -\frac{1}{4x^4} \times \frac{1}{x}\,dx =$$

$$-\frac{1}{4x^4}lnx + \int \frac{1}{4x^5}\,dx =$$

$$\int \frac{1}{4x^5}\,dx = \frac{1}{4}\int \frac{1}{x^5}\,dx = \frac{1}{4}\int x^{-5}\,dx = \frac{1}{4} \times \frac{x^{-4}}{-4} + c$$

$$= \frac{x^{-4}}{-16} + c = -\frac{1}{16x^4} + c$$

The final answer: $-\frac{1}{4x^4}lnx - \frac{1}{16x^4} + c = -\frac{lnx}{4x^4} - \frac{1}{16x^4} + c$

At this stage we use common denominator:

$$= \frac{-4lnx-1}{16x^4} + c$$

At the end we factorize the negative sign in the numerator:

$$= -\frac{4lnx+1}{16x^4} + c$$

$65) \int 2xarctanx\,dx =$

We use fractional method:

$$arctanx = u \rightarrow \frac{1}{1+x^2}\,dx = du, 2x\,dx = dv \rightarrow 2 \times \frac{x^2}{2} = v$$

$$\rightarrow x^2 = v$$

$$\int u\,dv = uv - \int v\,du = arctanx \times x^2 - \int x^2 \times \frac{1}{1+x^2}\,dx$$

$$= x^2 arctanx - \int \frac{x^2}{1+x^2}\,dx$$

$$\int \frac{x^2}{1+x^2}\,dx =$$

Note: If we add and deduce a certain value no change is made in the original value:

Example: $4 + 1 - 1 = 4$

Example 2: $x^2 + 1 - 1 = x^2$

$= \int \frac{x^2+1-1}{x^2+1} dx = \int \frac{x^2+1}{x^2+1} dx - \int \frac{1}{x^2+1} dx = \int 1 \, dx - \int \frac{1}{x^2+1} dx =$

$x - arctanx + c$

The final answer: $x^2 arctanx - x + arctanx + c$

---

66) $\int x^2 arctanx \, dx =$

We use fractional method:

$$arctanx = u \rightarrow \frac{1}{1+x^2} dx = du, x^2 dx = dv \rightarrow \frac{1}{3} x^3 = v$$

$$\int udv = uv - \int vdu = arctanx \times \frac{1}{3} x^3$$

$$- \int \frac{1}{3} x^3 \times \frac{1}{1+x^2} dx$$

$$= \frac{1}{3} x^3 arctanx - \frac{1}{3} \int \frac{x^3}{1+x^2} dx =$$

$$\int \frac{x^3}{1+x^2} dx =$$

Note: If we add and deduce a certain value no change is made in the original value:

Example: $x^3 + x - x = x^3$

$$= \int \frac{x^3+x-x}{1+x^2} dx =$$

We know $1 + x^2$ is the common denominator so the fraction can be expanded:

$$= \int \frac{x^3+x}{1+x^2} dx - \int \frac{x}{1+x^2} dx =$$

$$\int \frac{x^3 + x}{1 + x^2} dx =$$

We factorize x in the numerator:

$$= \int \frac{x(x^2+1)}{x^2+1} dx = \int x \, dx = \frac{1}{2} x^2 + c$$

$$\int \frac{x}{1 + x^2} dx =$$

We use change of variable method:

$$1 + x^2 = u \to 2x \, dx = du \to x \, dx = \frac{1}{2} du$$

$$= \int \frac{\frac{1}{2} du}{u} = \frac{1}{2} \int \frac{du}{u} = \frac{1}{2} \ln|u| + c \to 1 + x^2 = u \to = \frac{1}{2} \ln|1 + x^2| + c$$

The final answer: $\frac{1}{3} x^3 arctanx - \frac{1}{6} x^2 + \frac{1}{6} \ln|1 + x^2| + c$

$67)i = \int e^x \sin(4x) \, dx =$

We use fractional method:

$$e^x = u \to e^x dx = du, \sin(4x) \, dx = dv \to -\frac{1}{4} \cos(4x) = v$$

$$\int u dv = uv - \int v du = e^x \times -\frac{1}{4} \cos(4x)$$

$$- \int -\frac{1}{4} \cos(4x) \times e^x \, dx$$

$$= -\frac{1}{4} e^x \cos(4x) + \int \frac{1}{4} e^x \cos(4x) \, dx$$

$$\int \frac{1}{4} e^x \cos(4x) \, dx = \frac{1}{4} \int e^x \cos(4x) \, dx =$$

We use the method of change of variables:

$$e^x = u \rightarrow e^x dx = du, \cos(4x)\, dx = dv \rightarrow \frac{1}{4}\sin(4x) = v$$

$$\int u\, dv = uv - \int v\, du = e^x \times \frac{1}{4}\sin(4x) - \int \frac{1}{4}\sin(4x)$$
$$\times\, e^x dx$$

$$= \frac{1}{4}e^x \sin(4x) - \int \frac{1}{4}e^x \sin(4x)\, dx =$$

Here we multiply the obtained values by $\frac{1}{4}$:

$$\frac{1}{16}e^x \sin(4x) - \frac{1}{16}\int e^x \sin(4x)\, dx =$$

As you can see after re integration again the original value of i was obtained thus it can be concluded that this is recursive integral, so we use the following procedure:

$$i = -\frac{1}{4}e^x \cos(4x) + \frac{1}{16}e^x \sin(4x) - \frac{1}{16}i =$$

$$i + \frac{1}{16}i = -\frac{1}{4}e^x \cos(4x) + \frac{1}{16}e^x \sin(4x)$$

$$\frac{i}{1} + \frac{i}{16} = \frac{e^x \sin(4x)}{16} - \frac{e^x \cos(4x)}{4} \rightarrow \frac{16i + i}{16}$$
$$= \frac{e^x \sin(4x) - 4e^x \cos(4x)}{16}$$

$$\rightarrow \frac{17i}{16} = \frac{e^x \sin(4x) - 4e^x \cos(4x)}{16}$$

In order to remove the denominator 16 we multiply both sides by 16.

$$17i = e^x \sin(4x) - 4e^x \cos(4x) \rightarrow i$$
$$= \frac{e^x \sin(4x) - 4e^x \cos(4x)}{17}$$

$68) i = \int e^{3x} \sin x \, dx =$

We use fractional method:

$e^{3x} = u \rightarrow 3e^{3x} dx = du, \sin x \, dx = dv \rightarrow -\cos x = v$

$$\int u dv = uv - \int v du = e^{3x} \times -\cos x - \int -\cos x \times 3e^{3x} dx$$

$$= -e^{3x} \cos x + 3 \int e^{3x} \cos x \, dx$$

$$\int e^{3x} \cos x \, dx =$$

We use the method of change of variables:

$e^{3x} = u \rightarrow 3e^{3x} dx = du, \cos x \, dx = dv \rightarrow \sin x = v$

$$\int u dv = uv - \int v du = e^{3x} \times \sin x - \int \sin x \times 3e^{3x} dx$$

$$= e^{3x} \sin x - 3 \int e^{ex} \sin x \, dx$$

As you can see after re integration again the original value was obtained thus it can be concluded that this is recursive integral, so we consider the initial integral as i and use the following procedure:

$i = -e^{3x} \cos x + 3e^{3x} \sin x - 9i$

$i + 9i = 3e^{3x} \sin x - e^{3x} \cos x$

$$10i = 3e^{3x} \sin x - e^{3x} \cos x \rightarrow i = \frac{3e^{3x} \sin x - e^{3x} \cos x}{10}$$

$69) \int x^2 e^{3x} dx =$

We use the table

	x^2 and its derivatives	e^{3x} and its integrates

		x^2 ↘	e^{3x}
+		$2x$ ↘	$\dfrac{1}{3}e^{3x}$
−		2 ↘	$\dfrac{1}{9}e^{3x}$
+		0	$\dfrac{1}{27}e^{3x}$
−			

As you can see every other cell is given + and − signs and each left cell is multiplied by the right lower one:

The final answer:$\dfrac{1}{3}x^2e^{3x} - \dfrac{1}{9} \times 2xe^{3x} + 2 \times \dfrac{1}{27}e^{3x} + c$

$=\dfrac{1}{3}x^2e^{3x} - \dfrac{2}{9}xe^{3x} + \dfrac{2}{27}e^{3x} + c$

We factorize e^{3x}:

$=e^{3x}\left(\dfrac{1}{3}x^2 - \dfrac{2}{9}x + \dfrac{2}{27}\right) + c$

---

70) $\int x^3 \cos x \, dx =$

We use the table

		x^3 and its derivatives	$\cos x$ and its integrates
+		x^3 ↘	$\cos x$
−		$3x^2$ ↘	$\sin x$
+		$6x$ ↘	$-\cos x$
−		6 ↘	$-\sin x$
		0	$\cos x$

As you can see every other cell is given + and − signs and each left cell is multiplied by the right lower one:

The final answer: $x^3 sinx + 3x^2 cosx - 6xsinx - 6cosx + c$

---

71) $\int \frac{1}{1-cos2x} dx =$

Note: $cos2x = 1 - 2sin^2x$

$= \int \frac{1}{1-(1-2sin^2x)} dx = \int \frac{1}{1-1+2sin^2x} dx = \int \frac{1}{2sin^2x} dx =$

$\frac{1}{2} \int \frac{1}{sin^2x} dx =$

Note: $\frac{1}{sinx} = cscx \rightarrow \frac{1}{sin^2x} = csc^2x, \frac{1}{cosx} = secx \rightarrow \frac{1}{cos^2x} = sec^2x$

$= \frac{1}{2} \int csc^2x\, dx = \frac{1}{2} \times -cotx + c = -\frac{1}{2}cotx + c$

---

72) $\int \frac{dx}{1+sinx} =$

Note: We know that any number can be multiplied and then divided by a particular number without a change in its value:

Example ١: $4 \times \frac{5}{5} = \frac{4}{1} \times \frac{5}{5} = \frac{4\times5}{1\times5} = \frac{20}{5} = 4$

Example ٢: $\frac{1}{1+sinx} \times \frac{1-sinx}{1-sinx} = \frac{1-sinx}{(1+sinx)(1-sinx)} =$

The terms $1 - sinx$ are crossed out in the numerator and denominator

$= \frac{1}{1+sinx}$

Now we get back to the integral:

$\int \frac{1}{1+sinx} \times \frac{1-sinx}{1-sinx} dx = \frac{1-sinx}{(1+sinx)(1-sinx)} dx =$

The denominator is the sum of two squares, thus:

$$= \int \frac{1-sinx}{1^2-sin^2x} dx = \int \frac{1-sinx}{1-sin^2x} dx =$$

$Note: sin^2x + cos^2x = 1 \rightarrow cos^2x = 1 - sin^2x$

$$= \int \frac{1-sinx}{cos^2x} dx =$$

cos^2x is the common denominator thus the fraction can be expanded:

$$= \int \frac{1}{cos^2x} dx - \int \frac{sinx}{cos^2x} dx =$$

$Note: \frac{1}{cos^2x} = 1 + tan^2x = sec^2x$

$$= \int sec^2 x \, dx - \int \frac{1}{cosx} \times \frac{sinx}{cosx} dx =$$

$Note: \frac{1}{cosx} = secx, \frac{sinx}{cosx} = tanx$

$$= tanx - \int secxtanx \, dx = tanx - secx + c = \frac{sinx}{cosx} - \frac{1}{cosx} +$$

$$c = \frac{sinx-1}{cosx} + c$$

$$= \frac{sinx-1}{cosx} \times \frac{cosx}{cosx} = \frac{((sinx-1)cosx}{cos^2x} = \frac{(sinx-1)cosx}{1-sin^2x} =$$

The denominator is the sum of two squares

$$\frac{(sinx - 1)cosx}{(1 - sinx)(1 + sinx)} =$$

We factorize -1 in the numerator:

$$= \frac{-(1-sinx)cosx}{(1-sinx)(1+sinx)} =$$

$1 - sinx$ Are crossed out in the numerator and denominator

$$= \frac{-cosx}{1+sinx} + c$$

---

73) $\int tan^4x \, dx =$

Always keep in mind the following important formula:

$$\boxed{\int tan^n x \, dx = \frac{tan^{n-1}x}{n-1} - \int tan^{n-2}x \, dx}$$

$$= \frac{tan^3x}{3} - \int tan^2 x \, dx = \frac{1}{3}tan^3x - \int tan^2 x \, dx$$

$$\int tan^2 x \, dx =$$

$Note: tan^2x = sec^2x - 1 \rightarrow proof \rightarrow \frac{1}{cos^2x} - \frac{cos^2x}{cos^2x} =$

$\frac{1-cos^2x}{cos^2x} = \frac{sin^2x}{cos^2x} = tan^2x$

$\rightarrow \int tan^2 x \, dx = \int sec^2 x - 1 \, dx = \int sec^2 x \, dx -$
$\int 1 \, dx = tanx - x + c$

The final answer: $\frac{1}{3}tan^3x - tanx + x + c$

---

74) $\int cot^6 x \, dx =$

Always keep in mind the following important formula:

$$\boxed{\int cot^n x \, dx = -\frac{cot^{n-1}x}{n-1} - \int cot^{n-2}x \, dx}$$

$$= -\frac{cot^5x}{5} - \int cot^4 x \, dx = -\frac{1}{5}cot^5x - \int cot^4 x \, dx$$

$$\int cot^4x \, dx = -\frac{cot^3x}{3} - \int cot^2 x \, dx$$

$$= -\frac{1}{3}cot^3x - \int cot^2 x \, dx$$

$$\int \cot^2 x \, dx =$$

$Note{:}\cot^2 x = \csc^2 x - 1 \rightarrow$ proof $\rightarrow \dfrac{1}{\sin^2 x} - \dfrac{\sin^2 x}{\sin^2 x} = \dfrac{1-\sin^2 x}{\sin^2 x} =$

$\dfrac{\cos^2 x}{\sin^2 x} = \cot^2 x$

$= \int \csc^2 x - 1 \, dx = \int \csc^2 x \, dx - \int 1 \, dx = -\cot x - x + c$

$The\ final\ answer{:}\ -\dfrac{1}{5}\cot^5 x + \dfrac{1}{3}\cot^3 x + (-\cot x - x) + c$

$= -\dfrac{1}{5}\cot^5 x + \dfrac{1}{3}\cot^3 x - \cot x - x + c$

$75)\int \csc x \, dx =$

Note: We know that any number can be multiplied and then divided by a particular number without a change in its value:

$= \int \dfrac{\csc x}{1} \times \dfrac{\csc x - \cot x}{\csc x - \cot x} = \int \dfrac{\csc^2 x - \csc x \cot x}{\csc x - \cot x} \, dx =$

We use the method of change of variables:

$\csc x - \cot x = u \rightarrow -\csc x \cot x + \csc^2 x = du$
$$\rightarrow \csc^2 x - \csc x \cot x \, dx = du$$

$Note{:}y = \csc x \rightarrow y' = -\csc x \cot x, y = \cot x \rightarrow y' =$
$-(1 + \cot^2 x) = -\csc^2 x$

$\rightarrow y = -\cot x \rightarrow y' = - \times -\csc^2 x = +\csc^2 x$

$= \int \dfrac{du}{u} = \ln|u| + c \rightarrow \csc x - \cot x = u \rightarrow = \ln|\csc x - \cot x| +$

c

$76)\int \sec^5 x \, dx =$

Always keep in mind the following important formula:

$$\boxed{\int \sec^n x \, dx = \frac{\sec^{n-2} x \tan x}{n-1} + \frac{n-2}{n-1} \int \sec^{n-2} x \, dx}$$

$$= \frac{\sec^3 x \tan x}{4} + \frac{3}{4} \int \sec^3 x \, dx = \frac{1}{4} \sec^3 x \tan x + \frac{3}{4} \int \sec^3 x \, dx$$

$$\frac{3}{4} \int \sec^3 x \, dx = \frac{3}{4} \times \frac{\sec x \tan x}{2} + \frac{1}{2} \int \sec x \, dx = \frac{3}{8} \sec x \tan x$$

$$+ \frac{1}{2} \int \sec x \, dx$$

$$\int \sec x \, dx = \int \frac{\sec x}{1} \times \frac{\sec x + \tan x}{\sec x + \tan x} dx$$

$$= \frac{\sec^2 x + \sec x \tan x}{\sec x + \tan x} dx =$$

We use the method of change of variables:

$$\sec x + \tan x = u \rightarrow \sec x \tan x + \sec^2 x = du$$

$$Note: y = \tan x \rightarrow y' = 1 + \tan^2 x, 1 + \tan^2 x = \frac{1}{\cos^2 x} =$$

$$\sec^2 x$$

$$= \frac{1}{2} \int \frac{du}{u} = \frac{1}{2} \ln|u| + c \rightarrow \sec x + \tan x = u \rightarrow = \frac{1}{2} \ln|\sec x +$$

$$\tan x| + c$$

The   final   answer: $\frac{1}{4} \sec^3 x \tan x + \frac{3}{8} \sec x \tan x +$

$\frac{1}{2} \ln|\sec x + \tan x| + c$

---

77) $\int \sec^9 x \tan^5 x \, dx =$

$$\boxed{Reminder: a^n \times a^m = a^{n \times m}}$$

Example 1: $\sec^8 x \times \sec x = \sec^9 x$

Example 2: $\tan^4 x \times \tan x = \tan^5$

$$= \int \sec^8 x \sec x \tan^4 x \tan x \, dx = \int \sec^8 x \tan^4 x \sec x \tan x \, dx =$$

$$\boxed{\text{Note: } (a^n)^m = a^{n \times m}}$$

$Example:(tan^2x)^2 = tan^{2\times2}x = tan^4x$

$Note:tan^2x = sec^2x - 1$

$=\int sec^8x(sec^2x - 1)^2 \, tanxsecx \, dx =$

We use the method of change of variables:

$secx = u \rightarrow secxtanx \, dx = du$

$=\int u^8 (u^2 - 1)^2 du =$

$$\boxed{\text{Note: } (a - b)^2 = a^2 - 2ab + b^2}$$

$=\int u^8 (u^4 - 2u^2 + 1)du = \int u^{12} - 2u^{10} + u^8 du$

$=\int u^{12}du - 2\int u^{10} \, du + \int u^8 \, du = \frac{1}{13}u^{13} - 2 \times \frac{1}{11}u^{11} + \frac{1}{9}u^9 + c$

$=\frac{1}{13}u^{13} - \frac{2}{11}u^{11} + \frac{1}{9}u^9 + c$

$secx = u \rightarrow = \frac{1}{13}sec^{13}x - \frac{2}{11}sec^{11}x + \frac{1}{9}sec^9x + c$

---

78) $\int \frac{sin^7x}{cos^4x}dx = \int \frac{sin^6xsinx}{cos^4x}dx = \int \frac{(sin^2x)^3sinx}{cos^4x}dx =$
$\int \frac{(1-cos^2x)^3sinx}{cos^4x}dx =$

We use the method of change of variables:

$cosx = u \rightarrow -sinx \, dx = du \rightarrow sinx \, dx = -du$

$=\int \frac{(1-u^2)^3}{u^4} \times -du = -\int \frac{(1-u^2)^3}{u^4}du$

$$\boxed{\text{Note: } (a - b)^3 = a^3 - 3a^2b + 3ab^2 - b^3}$$

$$=-\int(1-3u^2+3u^4-u^6)u^{-4}\,du=-\int(u^{-4}-3u^{-2}+3-u^2)\,du$$

$$=-\int u^{-4}\,du+3\int u^{-2}\,du-3\int du+\int u^2\,du$$

$$=-\times\frac{u^{-3}}{-3}+3\times\frac{u^{-1}}{-1}-3u+\frac{u^3}{3}+c=\frac{1}{3u^3}-\frac{3}{u}-3u+\frac{1}{3}u^3+c$$

$$cosx=u\rightarrow=\frac{1}{3cos^3x}-\frac{3}{cosx}-3cosx+\frac{1}{3}cos^3x$$

$$Note:\frac{1}{cosx}=secx\rightarrow\frac{1}{cos^3x}=sec^3x\rightarrow=\frac{1}{3}sec^3x-3secx-3cosx+\frac{1}{3}cos^3x$$

---

79) $\int csc^7xcot^3x\,dx=\int csc^6\,xcscxcot^2xcotx\,dx=\int csc^6\,xcot^2xcscxcotx\,dx$

$Note:cot^2x=csc^2x-1$

$$=\int csc^6\,x(csc^2x-1)cscxcotx\,dx=$$

We use the method of change of variables:

$$cscx=u\rightarrow-cscxcotx\,dx=du\rightarrow cscxcotx\,dx=-du$$

$$=\int u^6\,(u^2-1)\times-du=-\int u^8-u^6du=-\int u^8\,du+\int u^6\,du$$

$$=-\frac{1}{9}u^9+\frac{1}{7}u^7+c\rightarrow cscx=u\rightarrow=-\frac{1}{9}csc^9x+\frac{1}{7}csc^7x+c$$

---

80) $\int\frac{3x+11}{x^2-x-6}dx=$

The denominator can be expanded by one common form, in this case we need two numbers the sum of which is -1 and their multiplication is -6, the numbers are -3 and 2:

$$\int \frac{3x + 11}{(x + 2)(x - 3)} dx =$$

Here we need to expand the fractions and then integrate the result thus: Fist draw two fraction lines and put a + sign between them put a and b in first and second numerator also put x+2 and x-3 in the first and second denominator. Then we do as follows:

$$\frac{a}{x + 2} + \frac{b}{x - 3} =$$

At this stage we use common denominator:

$$\frac{a(x - 3) + b(x + 2)}{(x + 2)(x - 3)} = \frac{ax - 3a + bx + 2b}{(x + 2)(x - 3)} =$$

We factorize x in the numerator:

$$= \frac{x(a+b)-3a+2b}{(x+2)(x-3)} =$$

Now we compare the numerator of the two fractions: X factor in the numerator of the original fraction is 3 then a+b=3. Also the fixed number in the numerator of the original fraction is 11. Then $-3a + 2b = 11$

$$\begin{cases} a + b = 3 \\ -3a + 2b = 11 \end{cases} \text{The } \textit{upper } \textit{line } \textit{in } \textit{multiplied } \textit{by}$$

$$3 \rightarrow \begin{cases} 3a + 3b = 9 \\ -3a + 2b = 11 \end{cases} \rightarrow 5b = 20 \rightarrow b = \frac{20}{5} = 4$$

$$a + b = 3 \rightarrow a + 4 = 3 \rightarrow a = 3 - 4 = -1$$

$$= \int \frac{-1}{x+2} + \frac{4}{x-3} dx = -\int \frac{1}{x+2} dx + 4 \int \frac{1}{x-3} dx$$

$$= -ln|x + 2| + 4ln|x - 3| + c = 4ln|x - 3| - ln|x + 2| + c$$

81) $\int \frac{dx}{16-x^2} =$

The first method: detailed

The denominator is the difference of square then:

$$= \int \frac{dx}{(4-x)(4+x)} =$$

We use the expansion of the fractions:

$$\frac{a}{4-x} + \frac{b}{4+x} = \frac{a(4+x) + b(4-x)}{(4-x)(4+x)} = \frac{4a + ax + 4b - bx}{(4-x)(4+x)}$$

$$=$$

We factorize x in the numerator:

Now we compare the numerator of the two fractions: X factor in the numerator of the original fraction dies no exist so we can assume that it is multiplied by 0 and removed thus a-b=0 also the fixed number in the numerator of the original fraction is 1 thus $4a+4b=1$. Now we place the obtained values in the system and solve the equations.

$$\begin{cases} a - b = 0 \\ 4a + 4b = 1 \end{cases} up\ line \times -4 \rightarrow \begin{cases} -4a + 4b = 0 \\ 4a + 4b = 1 \end{cases} \rightarrow 8b = 1$$

$$\rightarrow b = \frac{1}{8}$$

$$a - b = 0 \rightarrow a - \frac{1}{8} = 0 \rightarrow a = \frac{1}{8}$$

$$= \int \frac{\frac{1}{8}dx}{4-x} + \int \frac{\frac{1}{8}dx}{4+x} = \frac{1}{8}\int \frac{dx}{4-x} + \frac{1}{8}\int \frac{dx}{4+x} =$$

At this stage, each integral calculated separately and then summed up.

$$\frac{1}{8}\int \frac{dx}{4-x} =$$

We use the method of change of variables:

$$4 - x = u \rightarrow -dx = du \rightarrow dx = -du$$

$$=\frac{1}{8}\int \frac{-du}{u} = -\frac{1}{8}\int \frac{du}{u} = -\frac{1}{8}ln|u| + c = -\frac{1}{8}ln|4 - x| + c$$

$$\frac{1}{8}\int \frac{dx}{4+x} =$$

We use the method of change of variables again:

$$4 + x = u \rightarrow dx = du$$

$$=\frac{1}{8}\int \frac{du}{u} = \frac{1}{8}ln|u| + c = \frac{1}{8}ln|4 + x| + c$$

The final result: $\frac{1}{8}ln|4 + x| - \frac{1}{8}ln|4 - x| + c$

Second method: a quick method using the formula:

$$\int \frac{1}{a^2 - x^2} = \frac{1}{2a}ln\left|\frac{a+x}{a-x}\right| + c$$

$$\int \frac{1}{16 - x^2}dx = \int \frac{dx}{4^2 - x^2} =$$

$$a = 4 \rightarrow= \frac{1}{2 \times 4}ln\left|\frac{4+x}{4-x}\right| + c = \frac{1}{8}ln\left|\frac{4+x}{4-x}\right| + c$$

---

82) $\int_1^e \frac{lnx}{x}dx =$

As mentioned in chapter 1 to solve the problems of definite integral first we solve them as the indefinite integral and then obtain the upper and lower limits and deduct them.

$$\int \frac{lnx}{x} dx =$$

We use the method of change of variables:

At this point once we replace x by e and once by 1 and deduce the results.

$$=\frac{1}{2}(lne)^2 - \frac{1}{2}(ln1)^2 = \left(\frac{1}{2}\right) - (0) = \frac{1}{2}$$

Note: $lne = \log_e e = 1$

Conclusion: The logarithm of any number to base itself equals 1.

The logarithm of 1 to any base equals 0.

$$\left(\frac{1}{2}ln1\right)^2 = \left(\frac{1}{2} \times 0\right)^2 = (0)^2 = 0$$

---

83) $\int_0^2 \frac{e^x dx}{1+e^x} =$

First we have nothing to do with 0 and 2 and solve the integral as an indefinite one:

$$=\int \frac{e^x dx}{1+e^x} =$$

We use the method of change of variables:

$$1 + e^x = u \rightarrow e^x dx = du$$

$$=\int \frac{du}{u} = ln|u| \rightarrow 1 + e^x = u \rightarrow= ln|1 + e^x|$$

Now we obtain the upper and lower limit by replacing x with 2 and 0.

$$=ln|1 + e^2| - ln|1 + e^0| = ln|1 + e^2| - ln|1 + 1| = ln|1 + e^2| - ln2$$

We factorize ln:

$=ln[(1 + e^2) - 2]$

$Note: lnx - y = ln\frac{x}{y}$

$=ln\frac{1+e^2}{2}$

$84) \int_1^5 4^x dx =$

Always keep in mind the following important formula:

$$\boxed{\int a^x dx = \frac{a^x}{lna} + c}$$

$\rightarrow \int 4^x dx = \frac{4^x}{ln4}$

Now we obtain the upper and lower limit and deduce them by replacing x with 5 and 1.

$$=\frac{4^5}{ln4} - \frac{4^1}{ln4} = \frac{1024}{ln4} - \frac{4}{ln4} = \frac{1024-4}{ln4} = \frac{1020}{ln4}$$

$85) \int_0^1 \frac{\sqrt{x}}{1+x} dx =$

First we have nothing to do with 0 and 1 and solve the integral as an indefinite one:

$=\int \frac{\sqrt{x}}{1+x} dx =$

We use the method of change of variables:

$\sqrt{x} = u \rightarrow x = u^2 \rightarrow dx = 2u\,du$

$=\int \frac{u \times 2u\,du}{1+u^2} = 2\int \frac{u^2\,du}{1+u^2} =$

$Note: a + 1 - 1 = a$

Example: $u^2 + 1 - 1 = u^2$

$$=2 \int \frac{u^2+1-1}{1+u^2} du =$$

$1 + u^2$ is the common denominator so the fraction can be expanded:

$$=2 \int \frac{1+u^2}{1+u^2} du - 2 \int \frac{1}{1+u^2} du = 2 \int 1 \, du - 2 \int \frac{1}{1+u^2} du$$

Note: $\int \frac{1}{a^2+u^2} du = arctanu + c \rightarrow \int \frac{1}{1^2+u^2} du = arctanu$

$$=2u - 2arctanu \rightarrow \sqrt{x} = u \rightarrow = 2\sqrt{x} - 2arctan\sqrt{x}$$

Now we obtain the upper and lower limit and deduce them by replacing x with 1 and 0.

$$= (2\sqrt{1} - 2arctan\sqrt{1}) - (2\sqrt{0} - 2arctan\sqrt{0})$$
$$= (2 - 2arctan1) - (0 - 0)$$

Note: $arctan1$ means which angle's tan equals 1? Obviously, the answer will be $\frac{\pi}{4}$

$$=2 - \left(2 \times \frac{\pi}{4}\right) = 2 - \frac{\pi}{2}$$

---

86) $\int_{-1}^{1} e^{x^2} sin(x^9) dx =$

Note: If the function in front of integral is odd $\int_{-a}^{a} f(x) dx = 0$

An odd function is the one that if we replace x with –x, *f(x)is changed into - f(x). (y becomes –y)*

Example: $f(x) = x^3 - 2x$

$f(-x) = (-x)^3 - 2(-x) = -x^3 + 2x$

In case of the above problem the function is an odd one and the integration interval is -1 to 1 (-a, a) thus the Integral result will be 0.

$$\int_{-1}^{1} e^{x^2} \sin(x^9) dx = 0$$

87) $\int_0^{2\pi} \cos^2 x \, dx =$

First we have nothing to do with 0 and 2π and solve the integral as an indefinite one:

$$\boxed{\text{Note: } \cos^2 x = \frac{1 + \cos 2x}{2}}$$

$$= \int \frac{1+\cos 2x}{2} dx =$$

2 is the common denominator so the fraction can be expanded:

$$= \int \frac{1}{2} dx + \int \frac{\cos 2x}{2} = \frac{1}{2}x + \frac{1}{2}\int \cos 2x \, dx =$$

$$\boxed{\text{Note: } \int \cos ax \, dx = \frac{1}{a} \sin ax + c}$$

$$= \frac{1}{2}x + \frac{1}{2}\sin 2x$$

Now we obtain the upper and lower limit and deduce them by replacing x with 2π and 0.

$$= \left[\frac{1}{2}(2\pi) + \frac{1}{2}\sin(2\pi)\right] - \left[\frac{1}{2}(0) + \frac{1}{2}\sin(0)\right] = (\pi + 0) -$$
$$(0+0) = \pi - 0 = \pi$$

88) $\int_0^{\frac{\pi}{2}} \sin(4x) \cos(2x) \, dx =$

First we have nothing to do with 0 and $\frac{\pi}{2}$ and solve the integral as an indefinite one:

$$= \int \sin(4x)\cos(2x)\, dx =$$

> Note: $\int \sin(a)\cos(b)\, dx = \frac{1}{2}[\sin(a+b) + \sin(a-b)]$

$$= \int \frac{1}{2}[\sin(4x+2x) + \sin(4x-2x)] = \frac{1}{2}\int \sin(6x)\, dx + \frac{1}{2}\int \sin(2x)\, dx$$

> Note: $\int \sin(ax)\, dx = -\frac{1}{a}\cos(ax) + c$

$$= \frac{1}{2} \times -\frac{1}{6}\cos(6x) + \frac{1}{2} \times -\frac{1}{2}\cos(2x) = -\frac{1}{12}\cos(6x) - \frac{1}{4}\cos(2x)$$

Now we obtain the upper and lower limit and deduce them by replacing x with $\frac{\pi}{2}$ and 0.

$$= \left(-\frac{1}{12}\cos\left(6\frac{\pi}{2}\right) - \frac{1}{4}\cos\left(2\frac{\pi}{2}\right)\right) - \left(-\frac{1}{12}\cos(6 \times 0) - \frac{1}{4}\cos(2 \times 0)\right)$$

$$= \left(-\frac{1}{12} \times \cos(3\pi) - \frac{1}{4}\cos(\pi)\right) - \left(-\frac{1}{12}\cos(0) - \frac{1}{4}\cos(0)\right)$$

$$= \left(\left(-\frac{1}{12} \times -1\right) - \left(\frac{1}{4} \times -1\right)\right) - \left(\left(-\frac{1}{12} \times 1\right) - \left(\frac{1}{4} \times 1\right)\right)$$

$$= \left(\frac{1}{12} + \frac{1}{4}\right) - \left(-\frac{1}{12} - \frac{1}{4}\right) = \left(\frac{1}{12} + \frac{1}{4} + \frac{1}{12} + \frac{1}{4}\right) = \frac{1+3+1+3}{12} = \frac{8}{12} = \frac{2}{3}$$

---

89) $\int_0^{2\pi} \cos(5x)\cos(2x)\, dx =$

First we have nothing to do with 0 and 2π and solve the integral as an indefinite one:

$$= \int \cos(5x)\cos(2x)\, dx =$$

> Note: $\cos(a)\cos(b) = \dfrac{1}{2}[\cos(a+b) + \cos(a-b)]$

$$\cos(5x)\cos(2x) = \frac{1}{2}[\cos(5x+2x) + \cos(5x-2x)]$$

$$= \frac{1}{2}[\cos(7x) + \cos(3x)]$$

$$= \int \frac{1}{2}\cos(7x)\, dx + \int \frac{1}{2}\cos(3x)\, dx = \frac{1}{2}\int \cos(7x)\, dx + \frac{1}{2}\int \cos(3x)\, dx =$$

> Note: $\displaystyle\int \cos(ax)\, dx = \frac{1}{a}\sin(ax)$

$$= \frac{1}{2} \times \frac{1}{7}\sin(7x) + \frac{1}{2} \times \frac{1}{3}\sin(3x) = \frac{1}{14}\sin(7x) + \frac{1}{6}\sin(3x)$$

Now we obtain the upper and lower limit and deduce them by replacing x with 2π and 0.

$$= \left(\frac{1}{14}\sin(7 \times 2\pi) + \frac{1}{6}\sin(3 \times 2\pi)\right) - \left(\frac{1}{14}\sin(7 \times 0) + \frac{1}{6}\sin(3 \times 0)\right) =$$

$$\left(\frac{1}{14}\sin(14\pi) + \frac{1}{6}\sin(6\pi)\right) - \left(\frac{1}{14}\sin(0) + \frac{1}{6}\sin(0)\right) =$$

$$(\sin(\pi) + \sin(\pi)) - (0 + 0) = (0 - 0) = 0$$

$Note: \sin\frac{\pi}{2} = 1,\, sin\pi = 0,\, sin3\frac{\pi}{2} = -1,\, sin2\pi = 0$

$Note2 : \cos\frac{\pi}{2} = 0,\, cos\pi = -1,\, cos3\frac{\pi}{2} = 0,\, cos2\pi = 1$

90) $\int_1^2 \frac{x^5-x+5}{x^2} dx =$

First we have nothing to do with 1 and 2 and solve the integral as an indefinite one:

x^2 is the common denominator so the fraction can be expanded:

$$= \int \frac{x^5}{x^2} dx - \int \frac{x}{x^2} dx + \int \frac{5}{x^2} dx = \int x^3 dx - \int \frac{1}{x} dx + 5 \int \frac{1}{x^2} dx$$

$$= \frac{1}{4} x^4 - \ln x + 5 \int \frac{1}{x^2} dx$$

$$5 \int \frac{1}{x^2} dx = 5 \int x^{-2} dx = 5 \times \frac{x^{-1}}{-1} = -\frac{5}{x}$$

$$= \frac{1}{4} x^4 - \ln x - \frac{5}{x}$$

Now we obtain the upper and lower limit and deduce them by replacing x with 2 and 1.

$$= \left(\frac{1}{4}(2)^4 - \ln 2 - \frac{5}{2}\right) - \left(\frac{1}{4}(1)^4 - \square n1 - \frac{5}{1}\right) = \left(4 - \ln 2 - \frac{5}{2}\right) - \left(\frac{1}{4} - 0 - 5\right)$$

$$\rightarrow \frac{4}{1} - \frac{5}{2} = \frac{8-5}{2} = \frac{3}{2}$$

$$\rightarrow \frac{1}{4} - \frac{5}{1} = \frac{1-20}{4} = -\frac{19}{4}$$

$$= \left(\frac{3}{2} - \ln 2\right) - \left(-\frac{19}{4}\right) = \frac{3}{2} + \frac{19}{4} - \ln 2 = \frac{25}{4} - \ln 2$$

91) $\int_0^1 (\sqrt[3]{x} + \frac{1}{(1+x)^2}) dx =$

First we have nothing to do with 0 and 1 and solve the integral as an indefinite one:

$$= \int x^{\frac{1}{3}}\, dx + \int \frac{1}{(1+x)^2}\, dx = \frac{x^{\frac{4}{3}}}{\frac{4}{3}} + \int \frac{1}{(1+x)^2}\, dx = \frac{3}{4}x^{\frac{4}{3}} +$$

$$\int \frac{1}{(1+x)^2}\, dx = \frac{3}{4}\sqrt[3]{x^4} + \int \frac{1}{(1+x)^2}\, dx =$$

$$\int \frac{1}{(1+x)^2}\, dx =$$

We use the method of change of variables:

$$1 + x = u \rightarrow dx = du$$

$$= \int \frac{1}{u^2}\, du = \int u^{-2}\, du = \frac{u^{-1}}{-1} = -\frac{1}{u} \rightarrow 1 + x = u \rightarrow= -\frac{1}{1+x}$$

$$= \frac{3}{4}\sqrt[3]{x^4} + \left(-\frac{1}{1+x}\right) = \frac{3}{4}\sqrt[3]{x^4} - \frac{1}{1+x}$$

Now we obtain the upper and lower limit and deduce them by replacing x with 1 and 0.

$$= \left(\frac{3}{4} \times 1 - \frac{1}{2}\right) - (0 - 1) = \frac{1}{4} - (-1) = \frac{1}{4} + 1 = \frac{5}{4}$$

---

92) $\int_0^4 \log_4 64\, dx =$

First we have nothing to do with 0 and 4 and solve the integral as an indefinite one:

Note: $\log_a x^n = n \log_a x$

$$4^3 = 4 \times 4 \times 4 = 64 \rightarrow \log_4 4^3 = 3 \log_4 4 = 3 \times 1 = 3$$

$$= \int 3\, dx = 3x$$

Now we obtain the upper and lower limit and deduce them by replacing x with 4 and 0.

$$= (3(4)) - (3(0)) = 12 - 0 = 12$$

---

93) $\int_0^2 \frac{|x|}{x}\, dx =$

First we have nothing to do with 0 and 2 and solve the integral as an indefinite one:

Noting the absolute value:

The absolute value is presented as $|x|$ and defined as:

$$|x| = \begin{cases} x \rightarrow if \rightarrow x \geq 0 \\ -x \rightarrow if \rightarrow x < 0 \end{cases}$$

If the sign of the value inside the absolute value is positive it can be removed but if the value is negative we should multiply it by $-$ before removing the absolute value.

Example 1:$|7| = 7$

Example 2:$|-5| = - \times (-5) = +5$

Example 3:$|-x| = - \times (-x) = +x$

Note:$|x| = \sqrt{x^2}$

Example:$|cosx| = \sqrt{cos^2 x}$

As you can see in the above integral the range of 0-2 is positive and is always positive, so absolute mark can be easily eliminated

$= \int \frac{x}{x} dx = \int 1\, dx = x$

Now we obtain the upper and lower limit and deduce them by replacing x with 2 and 0.

$=2-0=2$

94)$\int_{-1}^{1} x^3\, |x| dx =$

Here the problem is divided into two parts the first part is 0 to -1 which is a negative range and it should be multiplied by -1 after removing the absolute mark and the second part is 0-1 which is a positive range and the absolute sign can be easily removed. Thus the integral is divided into two parts and after finding the value the results are added up.

1) $\int_{-1}^{0} x^3 (-x)dx = \int -x^4\, dx = -\frac{1}{5}x^5$

Now we obtain the upper and lower limit and deduce them by replacing x with -1 and 0.

Now we obtain the upper and lower limit and deduce them by replacing x with 1 and 0.

$=\left(\frac{1}{5}(1)^5 - \frac{1}{5}(0)^5\right) = \frac{1}{5}$

The final answer:$-\frac{1}{5} + \frac{1}{5} = 0$

---

95) $\int_0^1 \frac{dx}{x + \sqrt[3]{x}} =$

First we have nothing to do with 0 and 1 and solve the integral as an indefinite one:

In order to answer such problems with fractional exponents we use the change of variable $x = u^n$ in which n is the common denominator of the exponent of Xs.

$x^{\frac{1}{1}}, x^{\frac{1}{3}}$

We consider 3 as the common denominator of 1 and 3.

$$x = u^3 \rightarrow dx = 3u^2 du$$

$$x^{\frac{1}{3}} = (u^3)^{\frac{1}{3}} = u^{\frac{3}{3}} = u^1 = u$$

$$= \int \frac{3u^2 du}{u^3 + u} = 3 \int \frac{u^2 du}{u(u^2 + 1)} =$$

u^2 in the numerator is crossed out by u in the denominator.

$$3 \int \frac{u\, du}{1 + u^2} =$$

We use the method of change of variables:

$$1 + u^2 = z \rightarrow 2u\, du = dz \rightarrow u\, du = \frac{1}{2} dz$$

$$= 3 \int \frac{\frac{1}{2} dz}{z} = \frac{3}{2} \int \frac{dz}{z} = \frac{3}{2} \ln|z| \rightarrow 1 + u^2 = z \rightarrow = \frac{3}{2} \ln|1 + u^2|$$

$$u = x^{\frac{1}{3}} = \sqrt[3]{x} \rightarrow = \frac{3}{2} \ln|(\sqrt[3]{x})^2 + 1| = \frac{3}{2} \ln\left|\sqrt[3]{x^2} + 1\right|$$

Now we obtain the upper and lower limit and deduce them by replacing x with 1 and 0.

$$= \left(\frac{3}{2} \ln(\sqrt[3]{1^2} + 1)\right) - \left(\frac{3}{2} \ln(\sqrt[3]{0^2} + 1)\right) = \frac{3}{2} \ln 2 - \frac{3}{2} \ln 1 =$$

$$\frac{3}{2} \ln 2 - 0 = \frac{3}{2} \ln 2$$

Thanks to my dear mother and father

Dedicated to Mehdi Hashemi

The End